Nina Green

Climate change mitigation and adaptation strategies in African UNESCO biosphere reserves

Nina Green

Climate change mitigation and adaptation strategies in African UNESCO biosphere reserves

WÜRZBURG
UNIVERSITY
PRESS

Dissertation, Julius-Maximilians-Universität Würzburg
Philosophische Fakultät, 2023
Gutachter: Prof. Dr. Hubert Job, Prof. Dr. Susanne Becken

Impressum

Julius-Maximilians-Universität Würzburg
Würzburg University Press
Universitätsbibliothek Würzburg
Am Hubland
D-97074 Würzburg

www.wup.uni-wuerzburg.de
wup@uni-wuerzburg.de

2025 | Würzburg University Press
Print on Demand

Coverdesign: Holger Schilling
Titelbild: Nina Green

ISBN 978-3-95826-256-0 (print)
ISBN 978-3-95826-257-7 (online)
DOI 10.25972/WUP-978-3-95826-257-7
URN urn:nbn:de:bvb:20-opus-363792

Acknowledgements

The research journey I embarked on was transformative and enriching, and I would like to express my heartfelt gratitude to all who played a pivotal role in making it possible.

Academic Support and Guidance:

Professor Dr. Hubert Job, my Doktorvater, provided invaluable guidance and mentorship. I am eternally grateful for the opportunity to learn under your tutelage and for the challenges that broadened my perspective and made me a better researcher.

I also extend my heartfelt appreciation to Professor Dr. Susanne Becken, the secondary assessor of this thesis, for her insightful feedback.

Dr. Lutz Möller from the German Commission for UNESCO provided essential input and constant support throughout this study, for which I am deeply grateful.

I would also like to thank Univ.-Prof. Dr. Heiko Paeth and Univ.-Prof. Dr. Hartwig Frimmel for their mentorship, and Professors Dr. Ralf Klein and Dr. Manuel Woltering for their consistent support and assistance.

Finally, my sincere thanks go to Katrin Lang-Novikov, Dipl.-Ing (FH), for her exceptional work in creating informative maps that enhance this dissertation.

Peer Support and Friendship:

This research journey would not have been the same without the companionship and support of my fellow researchers at the Chair of Geography and Regional Research at the University of Würzburg: Dr. Lisa Majewski, Fidelcastor Frances Kimario, and Constantin Meyer. Your kindness, encouragement, and friendship have been invaluable.

I am especially grateful to Sarah Bittlingmaier and Anna Sophie Frieser for their support, friendship, and willingness to share in both the joys and challenges of this journey.

Financial Support and Research Participation:

I am deeply grateful for the financial support of The Oppenheimer Memorial Trust, which enabled me to pursue this research and have life-changing experiences.

I would like to thank the experts who provided feedback on my research instruments and participated in the pilot study, as well as all those who volunteered to participate in the questionnaire rounds. Your contributions were essential to the success of this research.

Personal Support and Love:
My deepest gratitude goes to my husband, whose unwavering love, encouragement, and support have been the cornerstone of this endeavour. Words cannot express how much your calming presence during moments of stress has meant to me.

I am eternally grateful to my parents, who have always been my rock and provided untiring support and guidance throughout my life. I appreciate your open ears for brainstorming ideas and your willingness to discuss my results. I am especially indebted to my mother, who played the crucial role of my editor throughout this process.

My brother, Marco, cousins Danelle and Heinrich, and aunt Madelein and uncle Andre Botha deserve special thanks for their love and support. Your spontaneous outings and warm hugs were a source of comfort and joy during this journey.

To my friends, both old and new, I am truly grateful for the shared coffee breaks, restaurant outings, festival experiences, and moments of joy. You have made this adventure even more memorable and enriched my life in countless ways.

This research journey has been a truly enriching experience, and I am deeply grateful to each and every one of you who played a role in making it possible. Your contributions will never be forgotten.

Table of Contents

List of Figures

List of Tables

Abbreviations

A	Actor
AfriMAB	African Biosphere Reserve Network
ArabMAB	Arab States Biosphere Reserve Network
BMZ	Federal Ministry of Economic Cooperation and Development (Germany)
CBD	Convention on Biological Diversity
CH_4	Methane
CIA	Central Intelligence Agency (US)
CNPPA	Commission on National Parks and Protected Areas
CO_2	Carbon dioxide
COP	Conference of the Parties
DRC	Democratic Republic of the Congo
ECO	Related Ecosystems
EU	European Union
FAO	Food and Agriculture Organization
G8	The Group of Eight
GDP	Gross Domestic Products
GIZ	German Agency for International Cooperation
GS	Governance Systems
$GtCO_2e$	Gigatons of carbon dioxide equivalent
IBM	International Business Machines Corporation
IGCP	International Gorilla Conservation Program
IHP	International Hydrological Programme
IPCC	Intergovernmental Panel on Climate Change
IUCN	International Union for Conservation of Nature
MAB	Man and the Biosphere Programme
MAB-ICC	MAB International Coordinating Council
MIT	Massachusetts Institute of Technology
N_2O	Nitrous Oxide
NOAA	National Oceanic and Atmospheric Administration
O	Outcome
O_3	Ozone
OECM	Other effective area-based conservation measures
Ppm	Parts per million
RS	Resource System

RU	Resource Unit
S	Social, Economic and Political Settings
SDG	Sustainable Development Goals
SPSS	Statistical Package for the Social Sciences
UK	United Kingdom
UN	United Nations
UNEP	United Nations Environment Programme
UNESCO	United Nations Educational, Scientific and Cultural Organization
UNFCCC	United Nations Framework Convention on Climate Change
US	United States of America
USD	United States Dollar
WAP	W-Arly-Pendjari Transboundary Biosphere Reserve
WMO	World Meteorological Organization

Abstract

This study aimed to investigate the biosphere reserves of the AfriMAB network to determine how they can collectively position themselves in their current institutional structures to better contribute to the CBD's 30X30 target and specifically improve their environment's resilience to climate change. This is important to help some of the most vulnerable countries mitigate and adapt to the impacts of climate change. Biosphere reserves are an ideal conduit for such activities as it is a landscape scale conservation initiative that includes both environmental conservation and sustainable development functions. However, because biosphere reserves' institutional structures differ, they each have a varying degree of success when implementing projects. This study compares the institutional structures of the AfriMAB network to determine which biosphere reserves are performing better and why, and where improvements can be made.

To achieve this, a modified version of the Ostrom (2009) social-ecological systems framework was employed, incorporating elements from the Delphi research method, and performance measures associated with the adaptive cycle. This framework was populated by relevant elements from key biosphere reserve governance documents relating to climate change, such as the "Statutory framework of the World Network of Biosphere Reserves," the "Lima Action Plan for UNESCO's Man and the Biosphere (MAB) Programme and its World Network of Biosphere Reserves (2016-2025)", the "Management Manual for UNESCO Biosphere Reserves in Africa", and "For Life, For the Future: Biosphere Reserves and Climate Change". This study adopted an empirical mixed-methods research approach, employing a cross-sectional research design. Data collection consisted of deploying two online questionnaires to the AfriMAB survey population, with a six-month interval between administrations. The initial survey took place from April 6 to June 15, 2022, and the follow-up was conducted from January 19 to March 30, 2023. Before distribution, the questionnaires underwent expert review and a pilot study. Non-probability sampling was the primary sampling method. Seventy respondents completed the first questionnaire, and 63 of them agreed to participate in the second round. This subset, now the new survey population, had active involvement of 38 respondents. Out of 34 (37%) and 35 (38%) AfriMAB Biosphere Reserves represented by these respondents in questionnaire rounds one and two, only 26 (28%) participated in both. Analysis of the collected data involved a combination of quantitative, qualitative, and spatial analysis techniques.

This study found that the UN must find ways to lower the cost of participation for members from less affluent countries to increase their involvement. The MAB programme must encourage its biosphere reserves to focus on the restoration and expansion of key biomes that will help society adapt to climate change. Biosphere reserves must conduct climate change impact studies and greenhouse gas assessments before they implement projects to improve their success rate. Climate change projects must commence at a rate of at least one project per three-year period, covering between 50-60% of the biosphere reserve, with an annual budget of at least USD 20 per km^2. The MAB programme must encourage its member states to acknowledge biosphere reserves in their legislation, and ideally provide financial support. There is also a need to improve the communication between member states' governance, biosphere reserves, and local communities.

These findings would hopefully provide valuable insights for managers and decision-makers, pinpointing areas of improvement, and facilitating informed actions. Additionally, the research addresses a knowledge gap concerning the current state of climate change initiatives within the AfriMAB network, particularly given its vulnerability to climate change impacts. A promising future avenue for research involves applying similar methodologies across all regional networks of the MAB programme. This approach would enable a thorough comparison of their performance in relation to climate change initiatives, offering valuable insights into regional variations and best practices.

Zusammenfassung

Ziel dieser Studie war es, die Biosphärenreservate des AfriMAB-Netzwerks daraufhin zu untersuchen, wie sie sich in ihren derzeitigen institutionellen Strukturen gemeinsam positionieren können, um einen wirksamen Beitrag zum 30X30-Ziel des CBD zu leisten und die Widerstandsfähigkeit ihrer Umwelt gegenüber dem Klimawandel zu verbessern. Dies ist wichtig, um einige der vom Klimawandel am stärksten gefährdeten Ländern dabei zu unterstützen, die Auswirkungen des Klimawandels abzumildern und sich an sie anzupassen. Biosphärenreservate sind ein ideales Instrument für solche Aktivitäten, da es sich um eine geschützte Natur- und Kulturlandschaft handelt, die sowohl Funktionen des Umweltschutzes als auch der nachhaltigen Entwicklung umfasst. Da sich die institutionellen Strukturen der Biosphärenreservate jedoch unterscheiden, sind sie bei der Umsetzung von Projekten unterschiedlich erfolgreich. Diese Studie vergleicht die institutionellen Strukturen des AfriMAB-Netzwerks, um zu ermitteln, welche Biosphärenreservate besser abschneiden, was die Gründe dafür sind und wo Potenzial für Verbesserungen besteht.

Um dies zu erreichen, wurde eine modifizierte Version des sozial-ökologischen Systemmodells von Ostrom (2009) verwendet, die Elemente der Delphi-Forschungsmethode und Leistungsmessungen im Zusammenhang mit dem Anpassungszyklus enthält. Dieser Analyse-Rahmen wurde durch relevante Elemente aus wichtigen Dokumenten der Biosphärenreservatsverwaltungen in Bezug auf den Klimawandel ergänzt, wie dem „Statutory Framework of the World Network of Biosphere Reserves“, dem „Lima Action Plan for UNESCO's Man and the Biosphere (MAB) Programme and its World Network of Biosphere Reserves (2016-2025)“, dem "Management Manual for UNESCO Biosphere Reserves in Africa" und dem "For Life, For the Future: Biosphärenreservate und Klimawandel". Diese Studie wählte einen empirischen Forschungsansatz mit gemischten Methoden und einem Querschnittsdesign. Die Datenerhebung bestand aus zwei Online-Fragebögen, die im Abstand von sechs Monaten an die Teilnehmer des AfriMAB-Netzwerks verschickt wurden. Die Erstbefragung fand vom 6. April bis 15. Juni 2022 statt, die Folgebefragung wurde vom 19. Januar bis 30. März 2023 durchgeführt. Im Vorfeld der Erhebung wurden die Fragebögen von Experten geprüft und einem Pre-Test unterzogen. Die primäre Methode zur Durchschnittbildung der Probandenzahl aus der Grundgesamtheit wird als Nicht-Wahrscheinlichkeitsstichprobe dargestellt. Siebzig Befragte füllten den Fragebogen in der ersten Erhebungsrunde aus, und 63 von ihnen erklärten sich zur Teilnahme an der zweiten Runde bereit, wovon letztlich 38 den Fragebogen der zweiten Erhebungsrunde beantworteten. Von den 34 (37%) bzw. 35 (38%) AfriMAB-Biosphärenreservaten, die von den

Befragten in der ersten und zweiten Fragebogenrunde vertreten wurden, haben nur 26 (28%) an beiden Runden teilgenommen. Die Analyse der gesammelten Daten umfasste eine Kombination aus quantitativen, qualitativen und räumlichen Analysetechniken.

Die Studie ergab, dass die Vereinten Nationen Wege finden müssen, um die Kosten für die Teilnahme von Mitgliedern aus weniger wohlhabenden Ländern zu senken, um deren Beteiligung am MAB-Programm zu erhöhen. Das MAB-Programm muss seine Biosphärenreservate dazu ermutigen, sich auf die Wiederherstellung und Ausweitung von regionalen Ökosystemen zu konzentrieren, die der Gesellschaft bei der Anpassung an den Klimawandel helfen werden. Die Biosphärenreservate müssen vermehrt Studien über die Auswirkungen des Klimawandels und die Bewertungen der Treibhausgase durchführen, bevor sie Projekte umsetzen, um ihre Erfolgsquote zu verbessern. Es sollte mindestens ein Projekt pro Dreijahreszeitraum gestartet werden, das 50 bis 60% der Fläche des Biosphärenreservats abdeckt, mit einem jährlichen Budget von minimum 20 USD pro km^2. Das MAB-Programm muss seine Mitgliedsstaaten ermutigen, Biosphärenreservate in ihrer Gesetzgebung anzuerkennen und idealerweise finanziell zu unterstützen. Außerdem muss die Kommunikation zwischen der Verwaltung der Mitgliedsstaaten, den Biosphärenreservaten und den lokalen Gemeinschaften verbessert werden.

Diese Ergebnisse können wertvolle Erkenntnisse für Manager und Entscheidungsträger liefern, Verbesserungsmöglichkeiten aufzeigen und fundierte Maßnahmen darlegen. Darüber hinaus schließt diese Studie eine Wissenslücke in Bezug auf den aktuellen Stand der Initiativen zum Klimawandel innerhalb des AfriMAB-Netzwerks. Ein vielversprechender zukünftiger Weg für die Forschung ist die Anwendung ähnlicher Methoden in allen regionalen Netzwerken des MAB-Programms. Dieser Ansatz würde einen Vergleich ihrer Leistungen in Bezug auf Klimawandel-Initiativen ermöglichen und wertvolle Einblicke in regionale Unterschiede sowie bewährte Maßnahmen bieten.

Résumé

Le but de cette étude était d'examiner les réserves de biosphère du réseau AfriMAB pour voir comment elles peuvent se positionner conjointement au sein de leurs structures institutionnelles actuelles pour mieux contribuer à l'objectif 30X30 de la CDB et améliorer la résilience de leur environnement au changement climatique. Cela est important pour aider certains des pays les plus vulnérables à atténuer les impacts du changement climatique et à s'y adapter. Les réserves de biosphère constituent un outil idéal pour de telles activités, car il s'agit d'une initiative de conservation du paysage qui inclut à la fois des fonctions de protection de l'environnement et de développement durable. Cependant, étant donné que les structures institutionnelles des réserves de biosphère diffèrent, elles connaissent des degrés de succès variables dans la mise en œuvre des projets. Cette étude compare les structures institutionnelles du réseau AfriMAB afin de déterminer quelles réserves de biosphère fonctionnent le mieux, pourquoi, et où des améliorations peuvent être apportées.

Pour y parvenir, une version modifiée du cadre des systèmes socio-écologiques d'Ostrom (2009) a été utilisée, incorporant des éléments de la méthode de recherche Delphi et des mesures de performance liées au cycle d'adaptation. Ce cadre a été complété par des éléments pertinents provenant de documents clés de gestion des réserves de biosphère liés au changement climatique, tels que le Cadre statutaire du Réseau mondial des réserves de biosphère, le «Plan d'action de Lima pour le Programme sur l'homme et la biosphère (MAB) de l'UNESCO et son Réseau mondial de réserves de biosphère. Réserves de biosphère (2016-2025) », le « Manuel de gestion des réserves de biosphère de l'UNESCO en Afrique » et « Pour la vie, pour l'avenir : les réserves de biosphère et le changement climatique ». Cette étude a adopté une approche de recherche empirique à méthodes mixtes et a utilisé une conception transversale. La collecte de données consistait en deux questionnaires en ligne envoyés aux participants à l'enquête AfriMAB à six mois d'intervalle. L'enquête initiale a eu lieu du 6 avril au 15 juin 2022 et l'enquête de suivi a été réalisée du 19 janvier au 30 mars 2023. Avant leur distribution, les questionnaires ont été examinés par des experts et soumis à une étude pilote. La principale méthode d'échantillonnage était l'échantillonnage non probabiliste. Soixante-dix répondants ont rempli le premier questionnaire et 63 d'entre eux ont accepté de participer au deuxième tour. Ce sous-groupe, qui représente désormais la nouvelle population interrogée, a été activement inclus par 38 répondants. Sur les 34 (37%) et 35 (38%) réserves de biosphère d'AfriMAB représentées par ces répondants dans les deux premières séries de questionnaires, seules 26 (28%) ont participé aux deux séries. L'analyse des données collectées comprenait une combinaison de techniques d'analyse quantitative, qualitative et spatiale.

L'étude révèle que l'ONU doit trouver des moyens pour réduire le coût de la participation des membres des pays les moins riches afin d'augmenter leur participation. Le programme MAB doit encourager ses réserves de biosphère à se concentrer sur la restauration et l'expansion des biomes clés qui aideront la société à s'adapter au changement climatique. Les réserves de biosphère doivent réaliser des études d'impact sur le changement climatique et des évaluations des gaz à effet de serre avant de mettre en œuvre des projets afin d'améliorer leur taux de réussite. Au moins un projet doit être lancé par période de trois ans, couvrant 50 à 60% de la réserve de biosphère, avec un budget annuel d'au moins 20 $ par km^2. Le programme MAB doit encourager ses États membres à reconnaître les réserves de biosphère dans leur législation et, idéalement, à les soutenir financièrement. De plus, la communication entre l'administration des États membres, les réserves de biosphère et les communautés locales doit être améliorée.

Nous espérons que ces résultats fourniront des informations précieuses aux gestionnaires et aux décideurs, mettront en évidence les opportunités d'amélioration et faciliteront une action éclairée. De plus, la recherche comble un manque de connaissances concernant l'état actuel des initiatives sur le changement climatique au sein du réseau AfriMAB, compte tenu notamment de sa vulnérabilité aux impacts du changement climatique. Une voie de recherche prometteuse réside dans l'application de méthodes similaires dans tous les réseaux régionaux du programme MAB. Cette approche permettrait une comparaison approfondie de leurs performances en matière d'initiatives sur le changement climatique et fournirait des informations précieuses sur les différences régionales et les meilleures pratiques.

1 Introduction

As an international challenge, climate change is silent, slow to develop, and ultimately lethal. It was observed for the first time in 1938 by Guy Callendar. He found that global temperatures were increasing and postulated that the most likely reason for this trend was the increasing emissions from economic and industrial activities (Fleming, 2013). This finding was widely disregarded until 1953, when Gilbert Plass proved that there is a connection between the increasing temperatures and industrialisation. He broadcast this discovery around the world, forcing meteorologists to recognise the impact of human activities on the atmosphere (Time, 1953). This realisation could not come at a better time, considering the world was just about to enter a new geological epoch, called the Anthropocene (Zalasiewicz et al., 2019).

For far too long, humanity has caused irrevocable and unprecedented changes to the Earth's surface, natural environment, and atmospheric composition, ushering in this new epoch that is shaped by human activities (Lewis & Maslin, 2015). The changes in the atmosphere gave rise to escalating temperatures, where each successive decade is warmer than the last, marked by a surge in hot days, heat waves, and prolonged periods of drought. The repercussions extend to the Arctic, where polar ice caps are melting. Elevated sea levels and modifications in sea ice patterns have adversely affected ocean circulation and the intricate mechanisms that govern global weather patterns. Consequently, more frequent, intense, and destructive storms became the norm, amplifying social and ecological challenges worldwide. This grim reality has devastating consequences for the biosphere, manifesting in a loss of biodiversity, widespread famine, heightened disease prevalence, and increased poverty (UN, 2023).

The most vulnerable societies to climate change are concentrated in sub-Saharan Africa. This is because many African societies rely on their country's natural resources, stable weather systems, and predictable rainfall to produce their livelihood, which is often climate-sensitive (e.g., agriculture or nature-based tourism) (Job et al., 2017; Schilling et al., 2020). Sub-Saharan Africa grapples with preexisting developmental challenges that significantly exacerbate the region's vulnerability to climate change. These challenges include issues of famine, overpopulation, political instability, intricate governance structures, and institutional dimensions (Job 2014; Tsomb et al., 2023). Additionally, there are constraints related to limited access to resources such as capital, markets, infrastructure, and technology. These interconnected factors underscore the complex landscape within which the impact of climate change unfold, and amplify the difficulties faced by the countries in this region (UNFCCC, 2020).

There are two fundamental ways to address the climate change challenge: either through mitigation or adaptation. Climate change mitigation relates to the reduction of greenhouse gas emissions and concentrations in the atmosphere. Climate change adaptation means minimising the impacts of climate change (Zhang & Ayyub, 2021). As Africa has relatively low greenhouse gas emissions, most climate change mitigation efforts go towards developing green economies to address future developmental needs. Green economies are characterised as being low-carbon, resource-efficient, and socially inclusive (Acar & Yeldan, 2019; Majewski & Job, 2019).

There is also a drive to address climate change through ecosystem-based adaptation. Ecosystem-based adaptation involves leveraging biodiversity and ecosystem services to help communities adjust to a more unpredictable climate. Ecosystem services comprise the direct and indirect contributions of ecosystems to human well-being, significantly influencing societies' survival and quality of life (Plieninger et al., 2016). These services can be classified into four categories: provisioning, regulating, cultural, and supporting services (e.g., absorption of carbon, flood regulation, water purification, temperature regulation, etc.) (Everard, 2021). In terms of climate change, there are numerous ways in which nature adapts and works to protect itself. For example, forests and wetlands act as natural filtrating reservoirs by drawing water from heavy rains into underground water supplies and storing it for times of drought. It has been found that in areas that are regenerating forests, groundwater supplies have risen by several meters (Ellison et al., 2017). Increasing the total tree cover further cools areas that are prone to heatwaves, not only through the shade the trees provide, but also by the release of water vapour through their leaves. Another example is biomes such as mangrove forests and coral reefs which ease the impact of severe floods and coastal storms by causing waves to break before they hit the shore. This lowers the force and height of their swell, which reduces the likelihood of the ocean degrading the coastline, especially with a rising sea level; for humans it means that these biomes protect settlements against the advancing ocean (UNEP, 2020). Investment in the management and restoration of these ecosystems is therefore essentially an investment in the well-being of the communities that rely on them (Kumar, 2022).

This is the logic behind the 30 by 30 (or 30X30) target set at the 2022 United Nations Biodiversity Conference (COP15), which aims to safeguard 30% of both land and oceans by the year 2030, with the overarching goal of enhancing the natural environment's capacity to absorb the impact of climate change (CBD, 2022a). Expanding the percentage of the natural environment under formal protection would ideally increase the environment's capacity to absorb more carbon from the atmosphere, and thereby act as a buffer against extreme weather and biodiversity loss, as well as safeguarding homes, crops, water supplies, and critical infrastructure (Stoll-Kleemann & Job, 2008; Lo & Jang, 2022).

One of the key avenues through which the international community attempts to achieve this is the International Union for Conservation of Nature's (IUCN) "other effective area-based conservation measures" (OECMs), which was introduced in 2010. It is defined by the Convention on Biological Diversity (CBD) as a: "geographically defined area, other than a protected area, which is governed and managed in ways that achieve positive and sustained long-term outcomes for the in-situ conservation of biodiversity, with associated ecosystem functions and services and where applicable, cultural, spiritual, socio-economic, and other locally relevant values" (CBD, 2018, p. 2).

While OECMs are a prominent option to achieve the 30X30 target, it is not as revolutionary as some would suggest (Dudley et al., 2018). UNESCO Biosphere Reserves have been recognising similar areas as part of their conservation model since 1996 (Price, 1996). The only difference now is that the IUCN has finally created a category for them, allowing areas that practice conservation as a secondary function, to be recognised and counted as part of the total global conserved area (Rodríguez-Rodríguez et al., 2021). UNESCO Biosphere Reserves will surely contribute to the 30X30 target by encouraging their members who fit the OECM description to apply for recognition. Biosphere reserves' potential involvement in climate change mitigation and adaptation could in itself be substantial (Barraclough & Måren, 2022), as they often include vast areas spanning hundreds or thousands of square kilometres. Through projects and collaboration with local communities, they conserve areas far beyond the boundaries of individual or formal protected areas, which gives them a significant reach and influence to expand the quality and quantity of conservation in the landscape (Reed & Price, 2019).

At its crux, the UNESCO Biosphere Reserves are learning areas where two seemingly opposing functions of nature conservation and societal development are reconciled (UNESCO, 2017). Biosphere reserves follow a three-zone conservation model, with core, buffer, and transition zones (Figure 1.1). The core zones contain formally protected areas for the conservation, research and monitoring of biodiversity and ecosystems. Surrounding it is the buffer zone where economic activities, social research and education take place; this is compatible with sound ecological practices and is comparable to OECMs. Lastly, there is a transition zone which takes the conservation of the landscape even further, as it aims to foster and investigate sustainable economic and social activities, such as tourism, agriculture, industry, and urban areas (UNESCO, 1995b).

Figure 1.1 UNESCO Biosphere Reserve model.

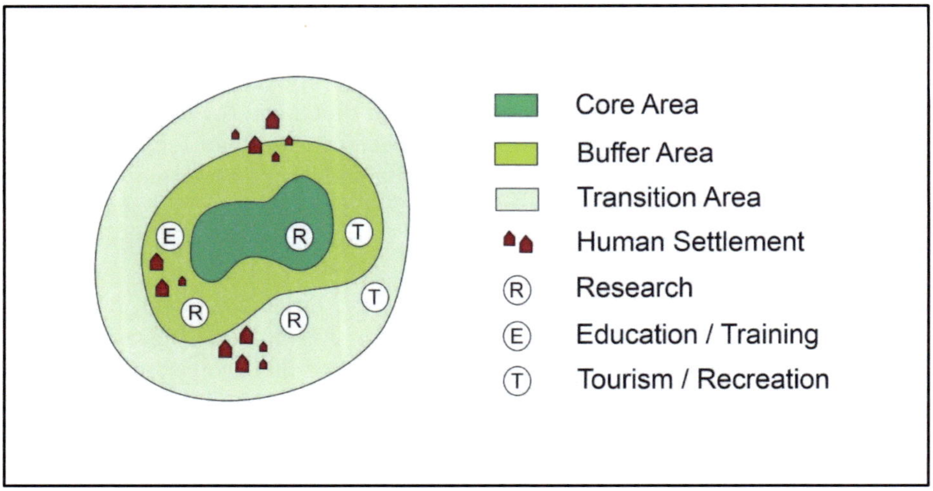

Source: Adapted from Chandrakar et al. (2016)

The biosphere reserve model therefore has to fulfil three main functions on a landscape scale, namely to conserve culture- and biodiversity; provide logistic support for monitoring, research, training and education; and foster environmental, cultural, and socially sustainable economic development (UNESCO, 2017). These areas are nominated and managed by their national governments but are designated by the Director-General under the intergovernmental Man and Biosphere (MAB) programme after the decision of the MAB International Coordinating Council (MAB-ICC). Currently, there are 748 UNESCO Biosphere Reserves in the World Network of Biosphere Reserves, covering almost 5% of the Earth's landmass in 134 countries. The network consists of five regional and seven thematic sub-networks. Africa has two networks: the Arab States Biosphere Reserve Network (ArabMAB) partly covering countries in northern Africa, and the African Biosphere Reserves Network (AfriMAB) that is in countries south of the Sahara Desert. The biosphere reserves are guided by the Statutory Framework of the World Network of Biosphere Reserves and its ten-year action plans (UNESCO, 2023). Climate change adaptation and mitigation activities feature within these guiding documents as they fall in line with the functions of biosphere reserves, making biosphere reserves theoretically an ideal conduit to address climate change.

1.1 Research problem

Biosphere reserves are individually managed by a variety of governance structures, which means that they all have different levels of success in accomplishing their main functions (Stoll-Kleemann et al., 2010). Each biosphere reserve therefore has a unique institutional structure, which is defined as a framework consisting of interconnected and interdependent social and ecological elements, forming a distinct arrangement within an organisation (Batie & Mercuro, 2008). These institutional structures are connected to each other through the regional and the World Networks of Biosphere Reserves.

For biosphere reserves to make a meaningful contribution to climate change adaptation and mitigation, and contribute to the 30X30 target, they must be able to respond effectively to new information and novel approaches to environmental management; they must be flexible and adaptable and, in extreme cases, be capable of changing their institutional (Walker & Salt, 2006). As early as the 1980s the MAB programme was already concerned about biosphere reserves' inability to adapt due to the rigid rules and regulations needed for intergovernmental organisations; these rules and regulations contradicted the adaptive management principles needed to manage individual biosphere reserves (Giorgio, 2003). An analysis of 156 biosphere reserves in 55 countries indicated that the primary reason for biosphere reserve failure stems from a lack of stakeholder participation due to these rigid policies and legislative (or top-down) reforms (Schultz et al., 2011). Adaptive management is key to promoting organisational learning, but its implementation is often constrained even on a local level because it involves risk and uncertainty (Giorgio, 2003). In African countries with limited conservation resources, this might be difficult. Organisations might start flexible with adaptive capacity but become institutionally rigid over time, which hampers their ability to readily adopt new approaches, strategies, and programmes (van Kerkhoff et al.; Gupta et al., 2010).

This phenomenon is described by the adaptive cycle metaphor of Holling (1986), which explains how the dynamics of social-ecological systems (such as biosphere reserves) change over time. The lack of effective adaptive management within these institutions is one of the biggest reasons why new approaches fail: often, these new approaches are simply 'stuck' on top of existing governance structures that were not designed to accommodate them (Walker & Salt, 2006). Also, the local governance framework is often unable to effectively implement the MAB action plan. This effect is observed in several of the original African biosphere reserves, where a biosphere reserve has been set up on top of a national park, or where a long-established reserve has been converted into a biosphere reserve (Botha et al., 2021). While biosphere reserves are struggling, the national parks that form their core areas are flourishing because they have a long-established governance structure and strategy, with more resources. These challenges ultimately affect the biosphere reserves' ability to implement climate change adaptation and mitigation strategies (Winkler, 2019). It is therefore important to compare the institutional structures of the different AfriMAB Biosphere Reserves, to determine why some biosphere reserves perform better than others and to identify areas of improvement.

1.2 The aim of this study

This study aims to investigate the biosphere reserves of the AfriMAB network to determine how they can collectively position themselves in their current institutional structures to better contribute to the CBD's 30X30 target and specifically to improve their resilience to climate change. To achieve this goal, several objectives were pursued, namely to:

1. Review relevant literature, broaden the theoretical understanding of how international climate change, nature conservation and MAB events influenced biosphere reserves' creation, evolution, and current institutional structures.
2. Determine which is the best theoretical framework to analyse biosphere reserves' complex institutional structures through a thorough literature review on systems thinking and social-ecological systems.
3. Establish the elements within the chosen social-ecological system framework that will be used as measures of each biosphere reserve's institutional structures, which are based on principal guiding documents for biosphere reserves.
4. Describe the methodology and methods used during this study that ensured that it was reliable, valid, and replicable and able to contribute to the academic body of knowledge on climate change and biosphere reserves.
5. Provide a brief introduction to the geographical and environmental context for the countries in sub-Sahara Africa and the African Regional Biosphere Reserve Network (AfriMAB).
6. Distribute a multi-round online questionnaire to 100 members of the AfriMAB network to determine what the current state of the biosphere reserves' climate change action is, and their performance.
7. Apply the chosen framework and make suggestions on how the AfriMAB Biosphere Reserve network can collectively position itself in its current institutional structures to better contribute to climate change adaptation and mitigation.

The significance of this research is that very little knowledge and literature exists about the current contribution of AfriMAB Biosphere Reserves to climate change adaptation and mitigation. This study intends to fill this knowledge gap, and by doing so, suggest steps that can be taken to enhance current climate change activities in one of the most vulnerable regions in the world. This will be achieved by investigating current structures and their shortcomings as well as opportunities for institutional improvements. In turn, this information could help biosphere reserve managers and decision-makers to address the identified issues and improve the overall institutional structures, and consequently enhance the networks' contribution to the 30X30 target.

1.3 Overview of chapters

The following chapter, chapter two, sets out to address objective one of this study, by reviewing the relevant literature on major climate change, nature conservation, and sustainable development events.

Chapter three goes on to address objectives two and three of this study. Initially, this chapter aims to identify the most suitable theoretical framework for analysing the intricate systems in this study. Subsequently, it intends to articulate the specific components of the institutional structures within the chosen framework that warrant investigation. This strategic approach will enable us to make a meaningful contribution towards achieving the overarching goal of advancing the AfriMAB network.

The following chapter, chapter four, delineates the methodology and employed methods in this study, ensuring its reliability, validity, and replicability. This chapter aims to achieve objective four of this study. By doing so, this research intends to contribute to the academic understanding of climate change actions within AfriMAB Biosphere Reserves.

In chapter five, the sub-Saharan African study area and AfriMAB network are introduced, providing a more specific context as to the reasons why this region was chosen for this study.

Chapter six presents the results of the two online questionnaires. These surveys were conducted over two distinct ten-week periods, with an interval of approximately six months between them.

Chapter seven concludes this study by restating the objectives and how they were achieved, followed by a summary of key suggestions on how the AfriMAB network can collectively position themselves in their current institutional structures to better contribute to the CBD's 30X30 target and specifically improve their environment's resilience to climate change. This chapter then goes on to discuss the limitations experienced during this study and suggest avenues for future research.

2 The influence of global events on the evolution of UNESCO Biosphere Reserves

2.1 Preliminary remarks

This chapter will review literature on discoveries and events relating to climate change, nature conservation, and sustainable development, that have influenced the establishment and evolution of UNESCO Biosphere Reserves. The objective of the chapter is therefore to discover: how have global events influenced biosphere reserves' creation, evolution, and current governance structures? This question will be answered by providing a brief overview of specific historical periods and elaborating on what the current UNESCO Biosphere Reserve management best practices and governance structures are.

2.2 Brief historical synopsis

This section will elaborate on how the following periods have influenced UNESCO Biosphere Reserves today: the Industrial Revolution (1750 to 1900); the first half of the 20th Century (1900-1950); the beginning of the environmental movement (1950-1990); turn of the Century (1990 to 2014); and the reinvigoration period (2015 to present).

2.2.1 Industrial Revolution (1750 to 1900)

Numerous climate and environmental issues we face today are rooted in the Industrial Revolution. It all started with the steam engine invented in 1712 by British ironmonger Thomas Newcomen, whose design was improved on and made more efficient by James Watt, a Scottish engineer, in 1776. These early engines were fuelled by coal or heavy fuel typically used for powering factories and transportation; this left many cities such as London under a constant industrial haze (Clapp, 2014; Kitsikopoulos, 2015).

Also during this time, weather measuring instruments became popularised and weather stations were established across the globe (Middleton, 1966). Key climate-measuring inventions predating this era were Evangelista Torricelli's mercury barometer (in 1643), Daniel Gabriel's Fahrenheit thermometer (in 1714), and Anders Celsius' standardised temperature scale (in 1742) (Manley, 1974; Teague & Gallicchio, 2017). This made it possible to accurately record temperatures; the oldest records are the Central England Temperature Record dating back to 1659 (Manley, 1974). The ability to measure and record atmospheric conditions accurately made it possible to start research on atmospheric conditions and climatic patterns (Le Treut et al., 2007). One seminal contributor was Joseph Fourier who published the paper titled "Remarques générales sur les températures du globe terrestre et des espaces planétaires" (General Remarks on the Temperature of the Terrestrial Globe and the Planetary Spaces) in 1824. He hypothesised that the Earth's atmosphere covers the planet like a blanket, trapping heat and keeping the planet warmer than it would be without an atmosphere (Fourier, 2013).

Eunice Newton Foote, an American amateur scientist, proved this theory. Her experiments showed that changes in the types and number of atmospheric gases, including carbon dioxide (CO_2) and water vapour, in Earth's atmosphere, trap escaping infrared (heat) radiation, warming the atmosphere. Her findings were published in a paper, titled "Circumstances affecting the heat of the sun's rays", which was presented at a meeting of the American Association for the Advancement of Science (AAAS) on Aug. 23, 1856. Some controversy exists regarding Eunice's discovery and very often John Tyndall, an Irish physicist who conducted similar experiments in the mid-19th century, has been credited with this discovery (Provenzale & Kilgarriff, 2023). Tyndall published his paper in 1861, titled "On the absorption and radiation of heat by gases and vapours, and the physical connexion of radiation, absorption, and conduction". It demonstrated that certain gases, including CO_2, were effective at trapping heat in the atmosphere (Tyndall, 1861).

On a different note, the first national park, Yellowstone, was proclaimed in 1872 to protect its landscape from industrialisation and urbanisation, specifically for the enjoyment of natural and wild landscapes by future generations. The concept of national parks rapidly spread to other countries and continents (Gissibl et al., 2012; Job, 2022). As most national parks were in remote areas, away from major urban hubs, the invention of the first Motorwagen in 1886 (by Karl Benz) made it increasingly easier to visit these spaces. However, in terms of environmental damage, these automobiles not only emitted greenhouse gasses, but they also degraded and fractured the landscape with the roads they required (Tsivacou, 1993; Nixon, 2016).

During the last few years of the 19th century, even more environmental damage caused by industrialisation was discovered. In 1896 Swedish chemist Svante Arrhenius quantified the contribution of CO_2 to the "atmospheric blanket" in his paper titled "On the influence of carbonic acid in the air upon the temperature of the ground" (Arrhenius & Holden, 1897). He coined the term "greenhouse effect" and hypothesised that the industrial era coal burning emissions would enhance it, and predicted that this would have a positive impact (Lowe, 2016). The "greenhouse effect" is caused by gasses, namely CO_2, methane (CH_4), ozone (O_3), and nitrous oxides (N_2O), that retain solar radiation and trap heat in the atmosphere, warming the Earth's surface temperature from -19°C to the current 15°C (von Weizsäcker et al., 1999; Cassia et al., 2018). While this is an important and positive function for life on Earth, increasing concentrations of these gasses are causing global average temperatures to rise, changing climate and weather patterns. It is estimated that compared to the pre-industrial era, the atmosphere now contains 40% more CO_2, with more than half of this increase appearing since 1970 (National Research Council, 2011; The Royal Society, 2014). Interestingly, Arrhenius' predictive models on what the effect of a doubling of CO_2 emissions would be on the climate, published in 1896, was in the same range as modern estimates, give or take a few degrees Celsius (Lowe, 2016).

The period 1850-1900 is often used as a reference for "pre-industrial" global temperatures. This was when the atmospheric CO_2 concentration was around 280 parts per million (ppm), in comparison to today's 400 ppm (IPCC, 2013). Other pre-industrial era markers refer to the sea level, glacial extent, and ice mass. As little pre-industrial era data is available on these markers, proxy data is used to determine the anthropocentric impact on them. Proxy data involves the measurement and analysis of, for example, ice cores and tree rings (The Royal Society, 2014).

2.2.2 The first half of the 20th century (1900 to 1950)

At the beginning of the 20th Century the world was slowly waking up to the possibility that human activities could be negatively impacting the environment, however, this progress was slowed as society was distracted by the First (1914-1918) and Second (1939-1945) World War. Still, just before the First World War broke out, two important innovations entered the market. The first was the Model-T automobile created in 1908, which revolutionised the motor and manufacturing industry. Ford Motors would become the beacon of industrial progress and mass production, with their revolutionary assembly line manufacturing technique (Parker, 2020). The second important development was the start of commercial air travel in 1914. These advances added to the increase in global CO_2 emissions, pushing it to one billion tons per year in 1927 (Clarke, 2006; Bitetti et al., 2013; Apps & Kassulke, 2023).

Between the First and the Second World War some progress was made regarding both protected area development and climate change related discoveries. In 1933 the International Conference for Protection of Fauna and Flora was held in London. Here attendees tried to define what constitutes a protected area and four categories were set, namely: national parks; strict nature reserves; fauna and flora reserves; and reserves with prohibition of hunting and collecting. These categories were not specifically defined, just generally named (Gissibl et al., 2012)

As environmental consciousness increased, British engineer Guy Callendar in 1938 noticed an upward trend in climatic temperatures (Le Treut et al., 2007). Using records from 147 weather stations around the globe, he proved that temperatures have risen over the past century and concluded that the most likely reason for it was increased concentrations of CO_2 in the atmosphere. His work, which he called the "Callendar effect", was dismissed by other meteorologists (Fleming, 2013).

In 1942, during the Second World War, the Western Hemisphere Convention on Nature Protection and Wildlife Preservation recognised four types of protected areas, namely: national parks; national reserves; nature monuments; and strict wilderness reserves. The classification of these protected areas would eventually contribute to the modern definitions of protected areas (Dudley, 2008).

After the Second World War, the international community made every effort to maintain and promote international peace and security, and to promote cooperation to solve social, economic, and humanitarian problems. Consequently, the United Nations (UN) was officially established in October 1945 when its Charter was approved by the Soviet Union, China, France, the United States, and the United Kingdom, among other signatories (Sayward, 2017).

The United Nations Educational, Scientific and Cultural Organization (UNESCO) was created soon after, in November of the same year. Its constitution, however, only came into force in 1946. The original goal of UNESCO was to aid countries in rebuilding cultural, educational, and scientific sites (e.g., schools, libraries, and museums), that were destroyed during the war in Europe. The declaration was signed by 26 signatories, including three countries from sub-Saharan Africa, namely, Ethiopia, Liberia, and South Africa (Duedahl, 2016). Three years later, in 1948, the first global environmental union was established in the French town of Fontainebleau. Its goal was to connect civil society and governments to protect nature. We still know this organisation today as the IUCN (Dudley, 2008).

Regarding climate research, Gilbert Plass, a Canadian physicist, published a detailed climate analysis in the journal "Tellus", titled "The carbon dioxide theory of climatic change" in 1956. He determined that doubled CO_2 concentrations would increase global temperatures by 3-4°C (Plass, 1956). He first presented these findings in May 1953 at the American Geophysical Union meeting. There he stated that the increase in industrial activity was discharging CO_2 into the atmosphere and causing the average global temperature to rise. His unpublished findings were broadcast by radio and newspaper services around the world, because although it was common knowledge that the temperatures were increasing, this was the first time that the phenomenon was publicly ascribed to industrial actions rather than natural systems. He further postulated that the world's vegetation and oceans would not be able to absorb the volume of carbon being emitted (Time, 1953), but this would only be proven five years later.

2.2.3 The beginning of the environmental movement (1950 to 1990)

The last half of the 20th century was marked by trailblazing international events for climate change and conservation initiatives. Increasingly more evidence regarding environmental degradation because of human activities came to light. In reaction, societal environmental consciousness emerged and developed into an environmental movement that would shape conservation as we know it today (Rannikko, 1996).

One of the first important discoveries of this period was made by United States (US) oceanographer Roger Revelle and chemist Hans Suess in 1957. They proved that seawater cannot absorb the excess carbon emitted from industry, as many – including Plass (1953) – have suggested. Revelle stated that "human beings are now carrying out a large-scale geophysical experiment that could not have happened in the past nor be reproduced in the future" (Keller, 1999, p. 368). A year later, in 1958, Charles David Keeling, an American scientist, designed and developed state-of-the-art equipment to systematically measure atmospheric CO_2. The equipment was installed at Mauna Loa in Hawaii, and in Antarctica. Within four years the project provided the first clear and unequivocal proof that CO_2 concentrations were rising. The model that was designed to visualise this increase is called the Keeling curve (Keeling, 2008).

In international news, the UN Economic and Social Council for the first time recognised that national parks and other protected areas are important for the sustainable use of natural resources. The IUCN's newly created Protected Areas Commission, now known as the World Commission on Protected Areas, made a first draft of a list of protected areas around the world. This list was presented at the first World Conference on National Parks in Seattle in 1962 (Dudley et al., 2010).

In the same year (1962) the book *"Silent Spring"* by Rachel Carson was published. This book warned society about the impact humans have on the environment. Her main message was however on the over-use of pesticides, but this book is often credited for kick-starting the international environmental movement (Carson, 1962). This gained the attention of government leaders and by 1965, the US President's Science Advisory Committee warned that a greenhouse effect was a matter of "real concern". It was the first time a government agency acknowledged that there was a problem and that a solution had to be found (US President's Science Advisory Committee, 1965).

Between 1966 and 1970 several important steps were taken internationally regarding the evolution of protected areas and landscape conservation. In 1966 the IUCN published its second version of the UN protected areas list, specifying three categories: national parks, scientific reserves, and natural monuments (Carson, 1962). In 1968 UNESCO held the Intergovernmental Conference of Experts on the Scientific Basis for Rational Use and Conservation of the Resources of the Biosphere, in Paris, France. This conference was attended by 326 delegates from 63 nations, and they made 20 recommendations for the sustainable use of natural resources. These recommendations emphasised the need for further global interdisciplinary research, education, and training, at all education levels, on the interactions between society and the biosphere (UNESCO, 1968). The aim of this research was specifically to encourage the collection of climate related data and then to share their knowledge with environmental managers, to improve their efficacy. The Deputy Director General of UNESCO opened the inaugural meeting in 1968 by saying, "Modern science might help select and develop rational methods for the use of the resources of the biosphere while ensuring their conservation" (Bridgewater, 2016, p. 395). This was the first time a protected area zonation system was mentioned, but within the context of research: the core of the protected area was to be conserved, while destructive types of research would be carried out in the buffer. It was from these recommendations that the idea of the UNESCO MAB Programme was born. However, the MAB Programme would only be launched three years later (Franklin, 1977).

On another note, in 1969 the IUCN held a General Assembly in New Delhi, India, where they finally defined what a national park is, namely: "a relatively large area where one or several ecosystems are not materially altered by human exploitation and occupation" (Phillips, 2004, p. 6). They subsequently asked countries not to call areas "national parks" if they did not fit this description (Phillips, 2004).

Then in 1971, the MAB Programme was finally launched, specifically for research on social-ecological systems and structures in the biosphere. With their research they wanted determine how the biosphere and its resources influenced, and were influenced by, society; as well as what the state of education and information available on these matters was. The overarching aim of all these objectives was to find ways (through research) to improve the relationship between society and the environment (Batisse, 1982; UNESCO, 2014).

In 1972 two important conferences were held. The first was the UN Conference on the Human Environment held in Stockholm, Sweden. Here representatives of different countries met to discuss environmental problems such as chemical pollution, atomic bomb testing, and whaling. Climate change hardly featured in these discussions, but the United Nations Environment Programme (UNEP) was consequently formed and governments were asked to start investigating the possible impacts of energy-related emissions on the environment (UN, 1973; Cherian, 2015). Later that same year the IUCN held the second World Conference on National

Parks in Yellowstone, in the US. Here Dr Dasmann, a senior ecologist at IUCN, proposed a new list of protected area categories (Dudley, 2008):

- Protected Anthropological Areas (Natural Biotic Areas, Cultivated Landscapes, Sites of Special Interest)
- Protected Historical or Archaeological Areas (Archaeological Sites, Historical Sites)
- Protected Natural Areas (Strict Natural Areas, Managed Natural Areas, Wilderness Areas)
- Multiple Use Areas
- National Parks
- Related Protected Areas (Provincial Parks, Strict Nature Reserves, Managed Nature Reserves, National Forests and Related Multiple Use Reserves, Anthropological, Archaeological or Historical Reserves)

This was the first time the IUCN recognised protected areas that were not strictly for ecological conservation purposes. However, it was also at this conference that the IUCN was asked to better "define the various purposes for which protected areas are set aside; and develop suitable standards and nomenclature for such areas" (Elliott, 1974; Dudley, 2008).

Finally, in 1974, biosphere reserves were founded after a meeting between UNESCO, UNEP, and IUCN to set the mission, vision, and objectives for the programme. They agreed on three main goals for biosphere reserves: first, to conserve genetic and biological diversity and their integrity; second, to allow areas for ecological and environmental research; and third, to provide facilities for education and training (Reed & Massie, 2013). The previously mentioned zonation system of the reserves was also introduced: a well-protected core zone surrounded by a buffer zone, where land use would increase the further away you move from the core (UNESCO, 1974; Batisse, 1982; Bridgewater, 2016). Two years later, in 1976, the World Network of Biosphere Reserves was launched, which was organised into regional and sub-regional networks (Pool-Stanvliet, 2013; Reed & Massie, 2013). The objective was that conservation and natural resource use should complement rather than compete with each other; and that interdisciplinary research must be used to achieve this goal (UNESCO, 1974).

While biosphere reserves and their networks were being developed, another climate change milestone was reached in 1975, when the term "global warming" was coined in a paper published by US scientist Wallace Broecker in "Science", titled "Climatic change: Are we on the brink of a pronounced global warming?" (Broecker, 1975).

In 1978, in response to the new international conservation areas – UNESCO Biosphere Reserves, and the World Heritage Convention – a more comprehensive list of protected area types was published by the IUCN (1978). The IUCN's Commission on National Parks and Protected Areas (CNPPA) also stipulated which type of protected areas they would be responsible for. The new categories are (IUCN, 1978, p. 11):

Group A: Categories for which CNPPA will take special responsibility:
- I. Scientific reserve
- II. National park
- III. Natural monument/National landmark
- IV. Nature conservation reserve
- V. Protected landscape

Group B: Other categories of importance to IUCN, but not exclusively in the scope of CNPPA:
- VI. Resource reserve
- VII. Anthropological reserve
- VIII. Multiple-use management area

Group C: Categories that are part of international programmes:
1. Biosphere reserve
2. World heritage site (natural)

This new list attempted to:

- Demonstrate how national parks can be enhanced by incorporating other types of protected areas.
- Assist nations in creating management categories that suit their specific needs.
- Ensure that, regardless of the naming conventions used by different countries, a conservation area can be accurately identified and classified based on its actual management objectives.
- Address inconsistencies arising from variations in administrative, institutional, legal, and political mechanisms.
- Aid the IUCN in gathering and analysing data on protected areas, facilitating storage, retrieval, updates, and printing of such information.
- Provide the scientific community with improved access to valuable conservation data (IUCN, 1978; Dudley & Stolton, 2008).

These categories, unfortunately, led to only more confusion as they did not properly define the different protected area types; several terms were used to describe the ten protected area types; a single protected area could be more than one type; and lastly, they lacked a marine protected area category (Dudley & Stolton, 2008).

The following year in 1979, the first World Climate Conference was held in Geneva, Switzerland, hosted by the World Meteorological Organization (WMO). This conference was ground-breaking in terms of climate change action. For the first time, nations were asked to try to prevent man-made changes to the atmosphere, as initial research showed that this would lead to climate change (Gaan, 2008). The result was the establishment of the World Climate Research Programme to further investigate this topic (Karling, 2001).

In 1980, the IUCN World Conservation Strategy was produced with the support of UNESCO, UNEP, and other organisations such as the World Wildlife Fund (WWF) and the Food and Agriculture Organization (FAO). This document was the first to address the use of living resource conservation and received input from governments, non-governmental organizations, and other experts. It was also the first time the concept of sustainable development was introduced, with the idea that development should support conservation and not deter it (Talbot, 1980). The term "sustainable development" was however only formally defined by the Brundlandt Commission in 1987 (Reed & Massie, 2013).

In 1984, the first International Biosphere Reserve Congress, hosted by UNESCO and UNEP in cooperation with the IUCN and FAO, convened in Minsk, Belarus. These organisations all had a vested interest in biosphere reserves. UNEP supported the biosphere reserve concept as it formed an international conservation network that supported global environmental monitoring. Already in 1984, there were 243 biosphere reserves in 65 countries (UNESCO, 1984; Brenner & Job, 2022). In Africa, there were 31 biosphere reserves that were mostly centred around the equator (Figure 2.1).

The IUCN, on the other hand, thought biosphere reserves were useful tools for regional planning and the promotion of the newly introduced concept of sustainable development, which biosphere reserves included quite early in its programme (Engelbauer, 2023). The FAO was interested in the biosphere reserve's potential to conserve genetic resources. The Minsk Congress resulted in the first guiding document for biosphere reserves, called the "Action plan for biosphere reserves". This document described the core conservation zone, surrounded by a single, or multiple buffer zones. The buffer zone was explained as an area of co-operation, where settlements, agriculture and other land uses would occur alongside conservation and research (UNESCO, 1984).

Figure 2.1 Biosphere reserve locations in 1984.

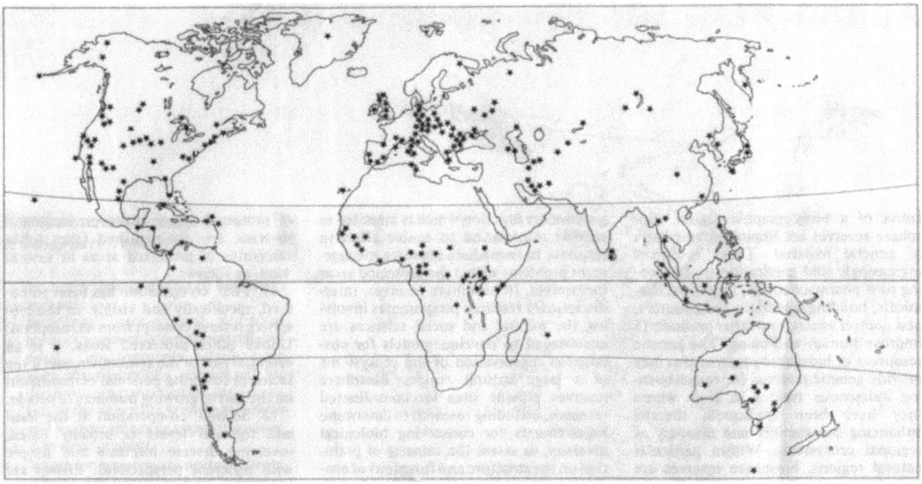

Source: UNESCO (1984, p. 4)

The minimum set of activities that biosphere reserves were required to fulfil was: "baseline inventories of flora and fauna and their uses; monitoring; preparation of a history of research; establishment of research facilities and research programmes; establishment of training and education programmes; and preparation of a management plan which addresses biosphere reserve functions" (UNESCO, 1984, p. 1). More specific responsibilities related to the climate goals set at the 1979 WMO conference that encouraged the global monitoring of the atmosphere and climate. In the action plan, it was stated that close collaboration with the WMO was needed as biosphere reserves have the potential to contribute to their monitoring goals. The suggested actions included two long-term monitoring activities: to identify parameters of global scientific significance, which were set to occur mainly between 1985 and 1986; and to use biosphere reserves for background monitoring of the atmosphere and long-term monitoring of the climate, mainly between 1986 and 1987. Both these activities were considered primary objectives in the original biosphere reserve action plan (UNESCO, 1984).

Unfortunately, these plans were disrupted in 1985 when the US, the United Kingdom (UK), and Singapore withdrew their membership from UNESCO. This caused the MAB programme to lose between 25%-30% of its operational budget and political support, which in turn crippled the MAB programme for several years, not only because of the loss of funds but also the loss of intellectual resources (Engelbauer, 2023). There were many explanations for these withdrawals, but the most common reasons related to UNESCO operational management, as it was considered to be inefficient, financially wasteful, and dangerously politicized (Hocking, 1985). Despite this controversy, Batisse (1985) encouraged governments and international organizations to enhance and expand the World Network of Biosphere Reserves. This included the advancement of essential knowledge for ecosystem conservation and biodiversity preservation, as well as improving the integration of conservation and development within biosphere reserves to fulfil MAB's overarching goals (Batisse, 1985).

While UNESCO and by extension the MAB Biosphere Reserve programme was suffering from its losses, organisations dealing with atmospheric and climate issues were making progress. In 1988 at the conference in Toronto, Canada, titled "Our changing atmosphere: Implications for global security", the Intergovernmental Panel on Climate Change (IPCC) was established (Taylor, 2014). The IPCC was endorsed by the UN General Assembly and tasked to prepare a comprehensive review of the state of knowledge and science regarding climate change; what the social and economic impact of climate change will be; and potential response strategies. That same year, the Helsinki Declaration for the Protection of the Ozone Layer (International Legal Materials, 1989) and the Montreal Protocol prohibiting the use of substances that diminish ozone, were published, which successfully led to the reduction of these harmful substances (Velders et al., 2007). The next year, in 1989, UK Prime Minister Margaret Thatcher called for a global treaty on climate change, as CO_2 emissions reached six billion tons per year (Le Treut et al., 2007).

2.2.4 The turn of the century (1990 to 2014)

At the turn of the century, the climate challenges escalated as more information became available about the extent of the problem, and international organisations increasingly tried to address them through international conferences and agreements. Biosphere reserves evolved and changed according to the needs of the international community.

This period started in 1990 with the publication of the IPCC's first assessment report which proved that global average temperatures rose between 0.3-0.6°C from the previous century and concluded that this was because of humanity's greenhouse gas emissions (IPCC, 1990). As a result, the UN General Assembly, a decision-making body within the UN, decided to convene in 1992 in Rio de Janeiro, Brazil, at the UN Conference on Environment and Development. Here a new framework for international agreements for the protection of the global environment was accepted, resulting in the Rio Declaration and Agenda 21. The overarching goal was to find ways in which greenhouse gas concentrations could be stabilized to prevent human activities from influencing the global climate system. For example, in Agenda 21, chapter 9, a link was established between human development, energy consumption, and the impact on the atmosphere. This turned out to be one of the most important climate change actions of its time, with 158 countries as signatories, which entered into force in 1994 (Zillman, 2009).

Once again, the issue of sustainable development was brought to the forefront of international policy, but this time in the context of climate change. During 1992, the MAB programme's ICC decided that the biosphere reserve concept needed attention, and an advisory committee was appointed (Bridgewater, 2016). By 1993 an internal report commented that despite the apparent focus on sustainable development in biosphere reserves on an international level, in practice people and social scientists studying them were underrepresented (Reed & Massie, 2013). The report also declared that like other protected areas, the designation and management of biosphere reserves excluded/marginalised locals from setting conservation objectives by not allowing them to identify appropriate local activities (in this case research), while also undermining opportunities for their sustainable development (Price, 1996). At this stage, the aim of biosphere reserves was still primarily to promote and encourage research and monitoring within and around core protected areas, creating living laboratories, with the purpose of improving its environmental management. These internationally designated areas mainly served scientific purposes. In 1994, even the IUCN decided to no longer classify biosphere reserves as "protected areas" because it did not align with their definition, namely "an area of land and/or sea especially dedicated to the protection and maintenance of biological diversity, and natural and associated cultural resources, and managed through legal or other effective means" (IUCN, 1994, p. 7). These categories included:

I. Strict protection
II. a) Nature reserve
III. b) Wilderness area
IV. Ecosystem conservation and protection (i.e., national park)
V. Conservation of natural features (i.e., natural monuments)
VI. Conservation through active management (i.e., habitat/species management area)
VII. Landscape/seascape conservation and recreation (i.e., protected landscape/seascape)
VIII. Sustainable use of natural resources (i.e., managed resource protected area)

Their argument for excluding biosphere reserves from this list was that biosphere reserves were international designations that were applied on top of already nationally designated IUCN-protected areas. At this stage, biosphere reserves were largely core zones with a small buffer area (IUCN, 1994). Some would argue that they were excluded because they included settlements and sustainable development goals, which did not align with the protected area definition (Engelbauer, 2023). However, the IUCN (1994) report stated that: "protected areas are not isolated units. Ecologically, economically, politically and culturally, they are linked to the areas around them" [...] "for the application of the categories system, however, where one area is used to buffer or surround another, both their categories should be separately identified and recorded" (IUCN, 1994, p. 13). This could have been possible regarding biosphere reserves because of category VI: Sustainable use of natural resources. These protected area categories implied a gradation of human intervention in protected areas (Dudley, 2008). However, this was a major setback for the visibility of MAB Biosphere Reserves (Engelbauer, 2023).

While UNESCO was reeling from the IUCN decision, 197 countries approved an international climate treaty which entered into force in 1994, called the UN Framework Convention on Climate Change (UNFCCC). The treaty aimed to investigate possible avenues to reduce global warming (mitigate) and to better address the inevitable temperature increase and its consequences (adapt) (Kuh, 2018). The following year, in 1995, the IPCC's second report again stated that humans were responsible for climate change (IPCC, 1995).

In 1995, the second UNESCO International Conference for Biosphere Reserves convened in Seville, Spain. At this stage, 324 biosphere reserves in 82 countries existed and 34 of them were in Africa (Brenner & Job, 2022; Engelbauer, 2023) (Figure 2.2).

At this conference, the original biosphere reserve concept was almost completely re-imagined according to the findings of the 1993 internal report and the UN's Agenda 21, emphasising the sociological aspect of biosphere reserves (UNESCO, 1995b). This conference delivered two documents that form the cornerstone of modern biosphere reserves, namely the Seville Strategy and the Statutory Framework for the World Network of Biosphere Reserves.

A conference paper was distributed to delegates that stated that the new purpose of a biosphere reserve was to "preserve and generate natural and cultural values through management that is scientifically correct, culturally creative and

Figure 2.2 Biosphere reserve locations in 1995.

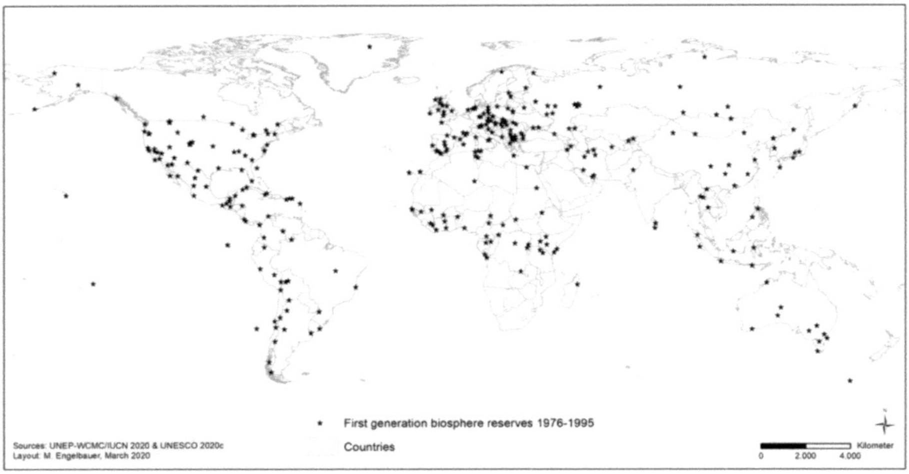

Source: Engelbauer (2023, p. 44)

operationally sustainable" (UNESCO, 1995a, p. 5). Biosphere reserves now had to fulfil three functions (UNESCO, 1995b):

- conserve biodiversity;
- foster sustainable social and economic development;
- support research, monitoring, and education.

This is also where the concept of the three-zone model was introduced, namely the core, buffer, and transition zones (Stoll-Kleemann & Welp, 2008). In terms of climate change, the strategy document specifically stated that biosphere reserves would make a greater contribution to the implementation of international agreements, and research programmes on climate change, desertification, and forests. However, unlike the original strategy document, they did not specify which organisations would be responsible for implementing and executing such activities (MAB, 1996).

In an article titled "People in biosphere reserves: An evolving concept", Price (1996) discusses two significant trends within biosphere reserves, namely sustainable development, and the increased acknowledgement of the importance of involving local communities in protected area management (Price, 1996). Biosphere reserves were not the only organisation to evolve. In 1995, the first Conference of the Parties (COP) was held; it is the decision-making body responsible for monitoring and revising the application of the UNFCCC (Arihant, 2021). Here the Berlin Mandate, which aimed to reduce emissions of industrialised countries, was adopted. It paved the way for the Kyoto Protocol by initiating talks on international, legal instruments, and stricter agreements. The subsequent protocol was officially adopted in 1997 and became the cornerstone of climate change actions because it required industrialized countries to reduce their emissions by at least 5% (below the 1990 levels) between 2008 and 2012. This came into force in 2007 with 160 signatories, but excluded the US that refused to ratify the treaty (von Weizsäcker et al., 1999). The results of this protocol were mixed; some signatories met their targets and others did not, but the lessons learned from it aided the 2015 UN Framework Convention on Climate, held in Paris, France (Rosen, 2015).

In 2000 UNESCO MAB held a five-year follow-up conference in Pamplona, Spain. Biosphere reserves were once again asked to focus more on socio-economic development along with their other goals. Challenges regarding the management of the new biosphere reserve format arose, as environmental managers who were previously in charge of the biosphere reserves did not have the authority to manage development or conservation activities outside the boundaries of the protected area, and often did not wish for the extra workload (UNESCO, 2000).

During 2000, the eight UN Millennium Development Goals were set to address the most pressing social, economic, and environmental issues of the 21st Century. Climate change fell under goal seven, namely, to promote environmental sustainability to reduce greenhouse gasses. All 191 UN member states, and 22 international organizations, signed this agreement (Ki-Moon, 2013). Regarding the international climate situation, in 2001, the IPCC in its third assessment report found that there is new and stronger evidence that greenhouse gas emissions from human activity cause the warming of the atmosphere (IPCC, 2001). US President George W. Bush subsequently removed his country from the Kyoto process (Ganapathy, 2021).

At the IUCN's 7th Conference of the Party (COP7), held in Kuala Lumpur in 2004, the organisation convinced the CBD and participating governments to accept its definition and categories of protected areas. This simplified protected area reporting and recording on an international level. The list however still did not include biosphere reserves (Dudley, 2008).

Between 2005 and 2008 there were several smaller climate change related milestones:

- In 2005, the Kyoto Protocol became international law for the countries affiliated with it, and the UK's Prime Minister, Tony Blair, chose climate change as a priority for his term as the chair of the G8 and president of the EU (Vogler & Bretherton, 2006).
- The Stern Review concluded that climate change could damage global GDP by 20%, but it would cost 1% of the global GDP to fix (Nordhaus, 2007).
- The same year carbon emissions reached eight billion tons per year (UNDP, 2007).
- The IPCC in their fourth assessment report in 2007 stated that 90% of greenhouse gas emissions were caused by human activities. This report also laid the groundwork for a post-Kyoto agreement, focusing on limiting warming to 2°C (Lemke et al., 2007).
- A Nobel Peace Prize was awarded to former US vice-president Al Gore for his efforts to reduce man-made greenhouse gas emissions (Rittberger et al., 2019).
- Then in 2008, after 50 years of observations at the Mauna Loa, the Keeling Project showed that the CO_2 concentrations had risen from 315 ppm in 1958, to 380 ppm in 2008 (Wilson, 2017).

Between 2007 and 2008 several management studies of the new biosphere reserve format were conducted. One study evaluated the perceptions of locals living near biosphere reserves and found there were conflicting opinions about biosphere reserves. Residents hoped that the biosphere reserve would improve regional development, but feared further restrictions on traditional land use (Wallner et al., 2007). In 2008, a study using three Czech biosphere reserves as models, indicated that the biosphere reserve concept can be used to effectively implement sustainable landscape ideas. However, it was found that complete integration of the biosphere reserves' environmental conservation and socio-economic development functions were deterred by formal and legislative constraints (Kušová et al., 2008).

The IUCN and UNESCO MAB Programme also had important conferences in 2008. The IUCN adjusted both the umbrella definition and categorisation of protected areas. The definition was changed to: "a clearly defined geographical space, recognised, dedicated and managed, through legal or other effective means, to achieve the long-term conservation of nature with associated ecosystem services and cultural values" (Dudley, 2008, p. 8). The categories were changed to:

Category Ia – strict nature reserve
Category Ib – wilderness area
Category II – national park

Category III – natural monument or feature
Category IV – habitat or species management area
Category V – protected landscape or seascape
Category VI – protected area with sustainable use of natural resources

These changes were adjusted to improve the definitions of IUCN-protected areas and clarify some of the more obscure categories, while ensuring that those categories do not get delisted because of the proposed changes (Dudley et al., 2010). Regarding UNESCO MAB Biosphere Reserves, in 2008 at the 3rd World Congress of Biosphere Reserves held in Spain, and the Madrid Action Plan which aimed to build on the Seville Strategy was activated for the period 2008-2013. It was found that only 23% of sites designated between 1976 and 1984 changed their biosphere reserve zoning system to the 1995 three-zone model (Price et al., 2010; Brenner & Job, 2022), which needed to be addressed.

This action plan also detailed the emerging challenge of climate change and the role(s) of the biosphere reserve network to address it (UNESCO, 2008), namely to:

- support international research programmes on climate change, especially because of its integrated approach to conservation and economic development;
- rapidly start monitoring, finding, and testing climate change solutions;
- become testing sites for demonstrating strategies for adaptation (in the core zone) and mitigation (in buffer and transition zoned);
- use forest and wetland biosphere reserves for carbon sequestration;
- build low-carbon economies through a mix of labour and technologically based social enterprises;
- support research into the political dimensions needed to cause societal lifestyle changes.

In 2011, UNESCO held the 40th Anniversary Conference of the MAB programme in Dresden, Germany, titled: "For life, for the future biosphere reserves and climate change". The overall goal of this conference was to encourage the MAB programme to give greater weight to climate change, which culminated in a declaration being signed by participants in support of the 1995 Seville Strategy and the 2008 Madrid Action Plan (Möller, 2011). At the policy level, member states were asked to agree to do the following (MAB, 2011):

- Emphasize the role of biosphere reserves in climate change mitigation and adaptation. Integrate their contributions into national and global climate policies.

- Increase the integration of biosphere reserves into development efforts. Support projects linking poverty eradication, biodiversity conservation, and climate protection in developing nations.
- Create appropriate legislative and administrative structures for biosphere reserves. Provide necessary funding, staff, and competencies for their management at the national and local levels.
- Develop biosphere reserves as models for sustainable development. Share successful practices widely.
- Back problem-focused, interdisciplinary research on climate change's impact on biosphere reserves, including traditional knowledge. Integrate findings into national and international research projects.
- Support economic activities combining climate change mitigation, ecosystem preservation, biodiversity, and social development. Address the needs of local communities, particularly in resource extraction and energy generation.
- Enhance the role of land use in carbon sequestration, especially forests. Implement and evaluate voluntary carbon Programme REDD+ in biosphere reserves to understand impacts, markets, and best practices.
- Facilitate clear communication about biosphere reserves and their processes to enhance public understanding.

Biosphere reserves were encouraged to (MAB, 2011):

- Develop and implement innovative climate change mitigation and adaptation strategies, integrating financing models and adjusting management plans. Strengthen regional resilience and sustainable development efforts.
- Formulate climate-responsive management plans based on vulnerability assessments. Promote biodiversity conservation; involve local communities; adapt to climate change.
- Combine traditional, indigenous, and local knowledge with modern scientific findings to enhance climate change research.
- Use biosphere reserves for sustainable development education. Stress how biodiversity conservation sustains ecosystem services and creates economic opportunities.
- Strengthen international collaboration by sharing experiences, effective methods, and joint projects through partnerships.
- Develop partnerships with the private sector to identify, develop, and promote sustainable products and services in local, national, and

international markets. Encourage climate-friendly economic practices in these sectors (MAB, 2011).

This conference was followed by a publication that showcased good climate change practices in biosphere reserves, which were considered to be ideal places to test, evaluate and implement comprehensive climate change policies (Möller, 2011).

In 2009 several important climate change related incidents occurred:

- China overtook the US as the biggest greenhouse gas emitter.
- Hackers downloaded emails from a server at the University of East Anglia's Climatic Research Unit which led to what was called "ClimateGate". These emails were misinterpreted by non-experts for the specific purpose of proving that global warming does not exist (Robinson, 2010).
- The UN Climate Summit held in Copenhagen was attended by representatives of 192 governments. It gave rise to the controversial Copenhagen Accord which excluded specific reduction targets from the final text, ensuring that it was not legal under international law (Mayer, 2022).

In 2010 more work was published commenting on biosphere reserve management globally. Research from all over the world indicated that biosphere reserves were undervalued and underutilised, and their roles and functions were not clearly understood by either government or the public (UNESCO, 2010b). Studies globally found that the top-down biodiversity approach which included a variety of stakeholders as well as community-led management, made biosphere reserve management a challenge (Stoll-Kleemann et al., 2010). A study by Schliep and Stoll-Kleemann (2010) supported these findings by emphasising the crucial role of national governance frameworks, professional knowledge, efficient communication, and active community engagement in biosphere reserve management. It also identified gaps in vertical integration among MAB institutions and national authorities, as well as shortages in trained personnel, funding, and local political backing for effective implementation (Schliep & Stoll-Kleemann, 2010).

Between 2010 and 2011 several smaller international climate change events took place. In 2010 countries began contributing to a $30 billion three-year deal to "fast start finance" to aid the greening of economies, and adapt to the impacts of climate change (Carfora & Scandurra, 2019). ClimateGate was reviewed by the IPCC and the scientists involved were cleared of malpractice, but the IPCC asked them for more openness. Then in 2011, in a new analysis of Earth's temperature, scientists concerned about the impact of ClimateGate proved that the Earth's land surface has warmed just as previous studies have shown (Maibach et al., 2012).

A study published in 2011 investigated the contribution biosphere reserves can make on social-ecological systems and learning for resilience. The researchers found that biosphere reserves offer practical insights into fostering social-ecological resilience. To maximize their role as learning hubs for sustainability, there's a need to enhance capacity and incentives for evaluating and communicating lessons learned (Schultz & Lundholm, 2010). A study by Schultz et al. (2011) on the impact of different stakeholders in biosphere reserves found that involving them and using adaptive co-management impacted management performance. In regular conservation, scientists' involvement helped, but volunteers' involvement hindered effectiveness. In sustainable development, residents' participation mattered greatly. The study indicated that using adaptive co-management made achieving development goals more effective, without harming biodiversity conservation efforts (Schultz et al., 2011).

The UN CBD held a conference in 2011 in Kanazawa City, Japan. Participants compiled the Strategic Plan for Biodiversity 2011-2020 and set the twenty global Aichi Biodiversity Targets. These targets were specifically created to address the growing rate of biodiversity loss. Two of these targets included climate change: under the sustainable goal section B: "reduce the direct pressure on biodiversity and promote sustainability", Target 10 aimed to minimise the impact of climate change on vulnerable ecosystems impacted by climate change. While target 15 set out to enhance the conservation and restoration of degraded ecosystems which would contribute to climate change mitigation and adaptation (CBD, 2011). The term OECM was introduced here for the first time, but it was not defined definitively – it was described as areas that achieve long-term and effective in-situ conservation of biodiversity outside of IUCN-defined protected areas (Dudley et al., 2018).

In 2011 a study on biosphere reserve management focusing on Mexico was published by Brenner and Job (2012), who also found that the contradicting nature of conservation-focused authorities and stakeholders wanting to use nature to support their livelihoods made effective management difficult, especially since the local population deeply mistrusts government authorities. This study reiterated the difficulty of finding a way to consolidate conservation with sustainable development outside protected areas (Brenner & Job, 2012).

In 2012, the UN Conference on Sustainable Development in Rio de Janeiro was held with the primary aim of addressing the urgent environmental, political and economic challenges of the world (Sachs, 2012). Between 2012 and 2013 the environmental situation continued to degrade. In 2012 Arctic Sea ice reached a record low summer coverage since satellite measurements first began in 1979 (Parkinson & Comiso, 2013). A biosphere reserve management paper titled: "Embracing ecological learning and social learning: UNESCO Biosphere Reserves as exemplars of changing conservation practices" showed that biosphere reserves had expanded its research focus to include the human dimension since 1995, but that the shift to incorporate the social dimension along with conservation made management more complicated (Reed & Massie, 2013). In 2013, a similar paper, "Reviewing biosphere reserves globally: Effective conservation action or bureaucratic label?", showed that the public still did not know or understand the term "biosphere reserve" (Coetzer et al., 2014).

Also in 2013, the Mauna Loa Observatory reported that the daily mean concentration of CO_2 in the atmosphere had surpassed 400 ppm for the first time since measurements began in 1958 (Showstack, 2013). In the IPCC's fifth assessment report, scientists stated that it is 95% certain that humans have been the "dominant cause" of global warming since the 1950s (IPCC, 2014).

In 2014 in a biosphere reserve management paper titled: "Biosphere reserves and their contribution to sustainable development: A value-chain analysis in the Rhön Biosphere Reserve, Germany" an effective solution for biosphere reserves to contribute to sustainable development was offered, namely through local product labelling. This enabled the biosphere reserve to educate local actors, especially business owners, about ecology while promoting a new economic mindset. These efforts aimed to reach the community and visitors as potential consumers. Effective communication, products, services, and the biosphere reserve label can thus positively impact the local economy (Kraus et al., 2014).

2.2.5 Reinvigoration (2015 to present)

The period before 2015 saw some controversy about the existence of climate change and whether humans were the cause of it. Biosphere reserves also experienced some uncertainties because of changes to the model. From 2015 however the international community showed a renewed interest in biosphere reserves as well as climate change. This renewed interest originated from the 2015 UN Climate Change Conference (COP21) in Paris where the so-called Paris Agreement was introduced. This document was a legally binding international treaty on climate change that was negotiated by 196 parties. The goal of this international treaty was to limit the global temperature increase from pre-industrial levels to below 2°C, but preferably 1.5°C, by 2100 (UNFCCC, 2016). The Paris Agreement pioneered a new approach to international climate policy as it united nations with a single cause: to mitigate climate change and adapt to its effects (Falkner, 2016). To achieve this goal, countries needed to reduce the estimated annual 52.7 $GtCO_2e$ (gigatons of CO_2 equivalent) of global greenhouse gas emissions present in the atmosphere in 2014, to 48 $GtCO_2e$ by 2025, and then to 42 $GtCO_2e$ by 2030. The objective was to achieve net-zero CO_2 emissions by 2060–75 (estimated at 35.5 GT in 2014), which is the balance between emission and reabsorption through carbon sinks (UNEP, 2015).

Limiting the average global temperature increase to 2°C as set by the Paris Agreement is expected to be a challenge, not to mention reaching the ideal target of keeping global temperatures below 1.5°C. Both depend on drastic carbon emission cuts (Henley & King, 2017). Without climate change intervention, it is estimated that global average temperatures will increase to above 4°C by 2050 (UNEP, 2015). Already, a temperature increase of 1.5°C is considered dangerous, and a 3°C increase catastrophic. The rising temperature in itself might not be as concerning, but the potential secondary effects are dangerous; this encouraged countries to participate in the Paris Agreement (Xu & Ramanathan, 2017). These effects include:

- *Extreme heat events*: A 1.5°C increase will cause mid-latitude regions to experience hot days that are 3°C warmer than in the pre-industrial era, with 14% of the global population exposed to extreme heat waves once every five years. A 2°C increase in the same regions will cause hot days to be 4°C warmer, with 37% of the global population experiencing extreme heat waves (IPCC, 2018; Dosio et al., 2018).
- *Rising sea levels*: Global sea levels, with a 1.5°C increase, are expected to rise with a median of 48-52 cm by 2100. This is expected to bring about an annual global flood damage adaptation cost of around USD 10.2 trillion. A 2°C increase is expected to cause a median sea level surge of 56-63 cm by 2100, causing a projected annual global flood damage and adaptation cost of USD 11.7 trillion. If these goals are not reached and no adaptations are made, it is expected that the median sea-level rise will be 0.86 m, which will cost USD 14.3 trillion to address (Jevrejeva et al., 2018; Rasmussen et al., 2018).
- *Declining biodiversity*: It is expected that 4% of terrestrial ecosystems will change with a 1.5°C temperature increase. It is projected that by 2100, 18-30% of plants, 11-28% of insects, and 4% of vertebrates will lose more than half of their habitats. A 2°C temperature increase will cause 13% of terrestrial ecosystems to change, resulting in the loss of more than half of the habitats of 23-38% of plants, 16-37% of insects, and 1-14% of vertebrates (IPCC, 2018; Warren et al., 2018).
- *Melting Arctic Sea ice*: If the temperature increases by 1.5°C it is projected to cause the Arctic Ocean to become ice-free in the summer once every 100 years. A 2°C increase would cause it to become ice-free every 10 years. This has negative implications for the oceans'

heat absorption, ocean currents, ecosystems, and weather systems (IPCC, 2018; Jahn, 2018).

- *At-risk coral reefs*: A 1.5°C increase is expected to cause a decline of between 70 and 90% of coral reef ecosystems. A 2°C increase is projected to cause a 99% loss (Frieler et al., 2012; Schleussner et al., 2016).
- *Declining global fisheries*: A 1.5°C increase will result in a decline of 1.5 million tons of fish and a 2°C increase will result in a decline of 3 million tons (Hoegh-Guldberg et al., 2018).
- *Rising poverty*: Societies dependent on natural, agricultural, or coastal resources for their livelihoods are particularly vulnerable to climate change. A 1.5°C increase instead of 2°C will reduce the number of people exposed to climate-related risk and poverty by hundreds of millions (Hallegatte & Rozenberg, 2017; Hoegh-Guldberg et al., 2018).
- *Health impacts*: Climate change is expected to cause approximately 250,000 additional deaths per year between 2030 and 2050 because of malnutrition; diseases like malaria, dengue fever, and diarrhoea; and heat stress. The projected direct damage cost to health is between US$2 and 4 billion per year by 2030 (WHO, 2014).
- *Food impacts*: Climate change without adaptation is likely to reduce global rice and wheat crops by 6% in the 2020s and 12%–15% by the 2080s (Aggarwal et al., 2019). It is expected that for every 1°C global temperature increase, wheat crops will decrease by 6% and rice yields by 10% (Moriarty & Honnery, 2018). Maize is expected to undergo a 9% crop loss in the 2020s and up to 20% by the 2080s (Aggarwal et al., 2019). This decrease is expected to particularly affect Southeast Asia, South America, and sub-Saharan Africa (Hoegh-Guldberg et al., 2018).

The UN also decided in 2015 to update the Millennium Development Goals to the Sustainable Development Goals for 2015-2030. This was a set of 17 targets. Goal 13 relates to climate change and the important actions needed to combat it (Sachs, 2012).

As mentioned earlier, biosphere reserves were also reinvigorated by the 2016-2025 Action Plan which was introduced in 2016 in Lima and succeeded the Madrid Action Plan. It updated the goals of the MAB Programme to include the 2030 Agenda for Sustainable Development and the Sustainable Development Goals which support the Paris Agreement (UNESCO, 2016). The Action Plan mentions climate change in three of its sections. First, the 2015-2025 MAB strategy introduction states that the MAB Programme and its members should support climate change adaptation and mitigation. Second, the vision and mission of the MAB programme proposes that the programme should support member states to explore and test technologies, policies, and innovations for sustainable environmental management. Third, the strategic action section suggests that biosphere reserves should be encouraged to become priority sites for climate change monitoring, research, adaptation, and mitigation. The performance of the biosphere reserves in addressing climate change would be measured by the number of climate change projects implemented, and whether the national government acknowledges the role of biosphere reserves in their climate change strategies (UNESCO, 2016).

In 2016, Reed wrote in the article "Conservation (in)action: Renewing the relevance of UNESCO Biosphere Reserves" that while the biosphere reserve network has supported research and practice, it remains underutilised by scientists and decision-makers. The article suggests that biosphere reserves should improve their collaboration with governments and conservation scientists to align with global initiatives like the CBD and Sustainable Development Goals. It advocates for establishing reliable transdisciplinary partnerships and engaging a wider range of experts to enhance conservation efforts (Reed, 2016).

An article on the management structures of biosphere reserves in Vietnam, (Cuong et al., 2017b) highlighted the difficulty of implementing biosphere reserves with a top-down approach. These biosphere reserves benefit from local autonomy despite weak national legal status, as administrative decentralisation allows local authorities to interpret policies that aid biosphere reserve management. Their sustainability in fact depends on locally tailored solutions (Cuong et al. (2017b). Writing about the challenges that biosphere reserves will face in the Anthropocene, Stoll-Kleemann and O`Riordan (2017) assert that for biosphere reserves to function as intended they must emphasise the role of local people, encouraging them to be pioneers for sustainable practices and to integrate ecological, social, and economic aspects of the area they live in (Stoll-Kleemann & O`Riordan, 2017). In research on the attributes of successful biosphere reserves published in 2018, Chong and colleagues found that such biosphere reserves belong to the post-Seville generation, and unsuccessful biosphere reserves tend to be non-compliant pre-Seville ones. They found the attributes of successful biosphere reserves were effective zoning, strong stakeholder participation, inclusive governance, adequate finance, and continuous monitoring for successful conservation and sustainable development (Cuong et al., 2018).

In 2018, IUCN officially defined OECMs as: "a geographically defined space, not recognised as a protected area, which is governed and managed over the long-term in ways that deliver the effective and enduring in-situ conservation of biodiversity, with associated ecosystem services and cultural and spiritual values" (IUCN, 2017, p. 11). OECMs come in three forms, namely: 1) areas whose primary objective is conservation in line with the IUCN protected area definition but are not officially recognised; 2) areas where conservation is a secondary management objective, such as watershed protection or climate change corridors; and 3) ancillary conservation areas where conservation outcomes are produced as a by-product of management activities (IUCN, 2017).

Areas cannot be defined as an OECM that are: 1) already recorded or part of an IUCN protected area and if it does not provide in-situ conservation of biodiversity; 2) does not have an essential conservation characteristic; and 3) when the conservation outcome is unsustainable. Therefore, biosphere reserves, covering a variety of land uses not contributing to the in-situ conservation of biodiversity and already including IUCN-protected areas, could not in its entirety be an OECM. However, sections within the biosphere reserve meeting these requirements can apply for an OECM status (UNESCO, 2022).

In a biosphere reserve management paper published in 2021 titled "Golija-Studenica Biosphere Reserve (Serbia) as a driver of change" the researchers described this biosphere reserve as being challenged by common global, biosphere reserve, and structural flaws such as a lack of supporting legislation; a centralised governing model; legally undefined biosphere reserve status; no regulation of stakeholder participation; and underfinancing (Pantić et al., 2021). In their paper on biosphere reserve management challenges, titled: "Potential and challenges of the Serengeti-Ngorongoro Biosphere Reserve, Tanzania", Botha et al. (2021) found that a lack of focus on the conservation of the socio-cultural heritage in the biosphere reserve has increased the social challenges affecting conservation, and that the biosphere reserve must therefore include modern OECM to address these challenges (Botha et al., 2021). Additionally, in their paper "Reviewing the participatory management of UNESCO Biosphere Reserves: What do we miss by ignoring local academic knowledge in Mexico?" Brenner et al. (2022) analysed research papers on Mexican biosphere reserves, and found that genuine stakeholder participation has not yet been achieved (Brenner & Job, 2022).

In November 2021, UN member states held a meeting in Glasgow, Scotland, to determine their progress in meeting the Paris Agreement targets. Coal use remained a concern, and the Glasgow Pact reaffirmed the need for the 1.5°C target set by the Paris Agreement (UNFCCC, 2021). A pledge of 100 billion dollars annually from developed to developing countries to build resilience to climate change and limit greenhouse gas emissions was welcomed as a step in the right direction

(Lennan & Morgera, 2022). The Glasgow Pact did however garner some criticism as some felt that the agreements made to reduce greenhouse gas emissions are not enough to curb the impact of climate change. They felt that a stricter, targeted, and more coordinated climate change governance regime is needed to effectively address climate change and limit the average global temperature increase to below 1.5°C (Masood & Tollefson, 2021).

In 2022 global emissions reached 55 $GtCO_2e$ (Forster et al., 2023). The Kunming-Montreal Global Biodiversity Framework was adopted during COP 15 in support of the sustainable development goals. This is where the 30X30 target was set that asked parties to: “ensure and enable that by 2030 at least 30 per cent of terrestrial, inland water, and coastal and marine areas are protected” (CBD, 2022b, p. 4). In terms of climate change, this target will increase the ability of the natural environment to absorb the impact of climate change. The most promising vehicle to accomplish this target is OECMs (Dudley & Stolton, 2022).

A biosphere reserve management paper published in 2022 by Paül et al., titled: “The n°1 country? A critical investigation of the booming designation of biosphere reserves in Spain” indicate that Spanish biosphere reserves were created for political and economic products and had very little in common with the UNESCO Biosphere Reserve goals (Paül et al., 2022).

The 2022 COP27 summit that was held in Sharm El-Sheikh was another important event in terms of climate change. It resulted in an implementation plan which included a commitment by richer nations to provide money to developing nations for damage and economic losses caused by the impact of ongoing climate change. The 200 participating nations were asked beforehand to submit plans to cut emissions by 2030 to prevent temperatures rising by more than 1.5°C (Allan et al., 2022). However, despite all these policies and efforts, the last recording of the Mauna Loa station in Hawaii showed that the CO_2 concentrations were at an all-time high in August 2023 at 419.68 ppm (NOAA, 2023).

A research article by Botha (2023) about biosphere reserve management titled: “Stakeholder perceptions of sustainability and possible behaviour in a biosphere reserve” recommends the adoption of a bottom-up management approach through the creation of sustainable tourism, environmental awareness, and increasing social cohesion amongst stakeholders.

The conclusion of 2023 marked the culmination of the UN climate summit (COP28) in Dubai, where the central agenda was to accelerate the global embrace of renewable energy, placing particular emphasis on the gradual elimination of fossil fuels. Notably, this conference was characterized by heightened tensions stemming from countries heavily reliant on fossil fuel production (Aljazeera, 2023).

Since the beginning, the MAB programme and biosphere reserves have been responsive towards international climate change and conservation developments and trends. Biosphere reserves therefore evolved conceptually over time and were even completely restructured in 1995, from research-focused protected areas into practicing conservation areas that include the entire landscape, not only the core protected areas. While this is a noble pursuit, research has shown that there were challenges in implementing such a concept practically on the ground, and even more so in biosphere reserves that lacked appropriate financial and governmental support, especially in less affluent countries. The following section will elaborate in more detail on the current governance structure of biosphere reserves and what is considered good governance strategies.

2.3 Biosphere reserve governance

Whilst UNESCO Biosphere Reserves are guided by the Statutory Framework and current Action Plans, their management is the responsibility of member states and the people whom they have appointed to operate the biosphere reserves. Unlike IUCN-protected areas, the role and legal status of biosphere reserves differ from country to country (Cuong et al., 2017b). Each biosphere reserve is unique and has individual financial mechanisms, approaches, designs, goals, objectives, and social-economic contexts (Mayer et al. 2018). Regrettably, not all these initiatives have been successful, as evidenced by failure examples all over the world (Wyborn, 2011).

To reduce the chances of such failure and to improve their ability to execute climate change mitigation and adaptation strategies effectively, biosphere reserves should aim to embody "good governance" principles that are "transparent, effective, and accountable (Carius & Job, 2019). It involves the whole population and considers the opinions and needs of minorities and the vulnerable" (BMZ, 2022, p. 1). However, good governance of biosphere reserves and subsequent climate change projects require a basic governance structure to function effectively, which includes elements such as a constitution, a legal structure, an environmental management plan, a business plan, and a supporting infrastructure (African Wildlife Foundation, 2016).

Once this basic governance structure is established, managers should operate the biosphere reserve following seven principles in their management strategies:

- Acknowledge and manage it knowing that the relationship between the natural environment and the people living in the landscape is complex, interactive, and unpredictable (Folke, 2006). It is thus recommended that managers follow an adaptive management strategy (Newsome et al., 2013).

- Cultivate a variety of strategies, contingencies, methods, and social and financial capital that will help the biosphere reserve deal with any problems that might occur. This will help managers and participants to react quickly and effectively to any situation (Simonsen et al., 2014; Job et al., 2023).
- Manage the connectivity within the landscape. There is a balance between well-connected landscapes that can help a system recover faster from a problem, and an overconnected initiative which may lead to the rapid spread of a problem, for example, a wildfire in a landscape without firebreaks (Folke et al., 2016).
- Include future problems which appear slowly, like climate change and desertification, in the environmental management plan. This will help to protect ecosystem services within the landscape in the long term (Simonsen et al., 2014).
- Encourage research, experimentation, innovation, and learning among the participants and external stakeholders. This will produce new knowledge in the biosphere reserve and create the opportunity for organisational learning (Folke et al., 2016);
- Increase stakeholder participation by promoting the following practices: Reduce the cost of stakeholder participation by keeping meetings short, specific, and only when necessary (Olsson et al., 2004; Job & Weizenegger, 2006); make the impact of participation more explicit by using local media to highlight success stories; and create an element of enrichment for stakeholders through training, expert advice, and knowledgeable speakers. These actions will help managers form a shared understanding of problems, build trust, and discover new perspectives (Holling, 2001; Simonsen et al., 2014). This will limit the free-rider problem, which occurs when individuals feel that their contribution to discussions are irrelevant, although they ultimately still benefit from decisions made in the meetings (Stroup, 2000).
- In most biosphere reserve networks, the state remains an important actor, along with international, government, subnational, parastatal, non-government, and private stakeholders. Managers must therefore work with these stakeholders to help connect the biosphere reserve in an effective polycentric governance system (Simonsen et al., 2014; Folke et al., 2016).

Most of the basic governance structures and strategies are possible to implement, even by less affluent biosphere reserves located in the global south. Still, management of UNESCO Biosphere Reserves and associated climate change projects under the UNESCO MAB Programme is complex and varies significantly from one biosphere reserve, country, and network, to the next.

2.4 Concluding remarks

In this chapter the objective was to determine how global events relating to climate change, nature conservation, and sustainable development have influenced biosphere reserves' creation, evolution, and current governance structures. To understand the present challenges, it is vital to understand the past. Climate change concerns emerged around the Industrial Revolution as temperature and atmosphere measurement and record-keeping became mainstream (Weart, 2008, p. 13). The results sparked a debate about the impact of human activities on the atmosphere and the global warming trend (May, 2020). Meanwhile, as more industries developed, the industrialising countries' population started migrating to the cities, especially as agriculture became heavily mechanised (Antrop, 2004). Natural landscapes were consequently converted into places supporting industry or urbanisation instead of biodiversity or ecosystem services. This ecological degradation, in turn, caused environmental conservation movements, with the rise of national parks and other protected areas. However, as time went by, these protected areas turned into islands of conservation in a sea of man-made landscapes (Gissibl et al., 2012). The juxtaposition between development and conservation needs became an international concern, and ultimately resulted in the creation of biosphere reserves to find a balance (McCormick, 1995). With every major international environmental and climate change conference, biosphere reserves evolved to incorporate the latest international targets. This created a complex system of biosphere reserves, each with unique management structures and strategies.

3 Theoretical Framework

3.1 Preliminary remarks

This chapter has two objectives which, when realised, will form the theoretical framework of this study. The first objective is to select the best theoretical framework for analysing this study's complex systems. The second objective is to clearly define the different elements[1] of the institutional structures contained in the selected framework. By employing these objectives, we can explore the research aim of finding ways for the AfriMAB network to improve both its resilience to climate change and its contribution to the CBD's 30X30 target, while operating within its current institutional structures.

The theoretical approaches that will be examined include systems thinking, social-ecological systems and the social-ecological framework, the self-fulfilling prophecy theory, and the Delphi research method performance measures, as well as the adaptive cycle.

3.2 Systems thinking

In this study, we will use systems thinking as a cognitive paradigm, as it assumes that reality is created from patterns that emerge when various elements interact (Randle & Stroink, 2018). The theoretical principles on which systems thinking is based originated in the 1940s from different authors in a variety of disciplines. Scientists recognised that the traditional paradigms used up until this point in history were inadequate to address the increasing complexity of social, economic, and environmental problems (Machlis et al., 1997). Academic research mostly followed the reductionist paradigm which, contrary to systems thinking, neglected relations and connections between elements and tended to analyse them in isolation (Bertalanffy, 1950). This paradigm was therefore often criticised for oversimplifying complex situations and real-world problems (Chowdhury, 2019, p. 6,7).

Norbert Wiener, an American mathematician who established the field of cybernetics, and whose work on control and communication within living organisms and machines, laid the theoretical foundations for systems thinking, that are still used today (Wiener, 1948). Another major contributor was Kurt Lewin, a German-American psychologist, who developed the theories and methods needed to research social systems, and who also determined that social connections affect behaviour (Lewin, 1951). Their work, while not directly mentioning systems thinking, provided some of the field's fundamental concepts (Ramage & Shipp, 2020).

---

1 An element is: "A component part of a complex whole". Oxford English Dictionary (2023, p. n.p.).

It was only when an American computer engineer and systems scientist, Jay Wright Forrester, founded the field of system dynamics at the Massachusetts Institute of Technology (MIT) that systems thinking as a single field of study started to emerge, as it provided a theoretical foundation for systems thinking (Forrester, 1999). Forrester defined a system as a set of interrelated components working together to achieve a common purpose (Forrester, 1961). In his first book, “Industrial dynamics” (1961), he applied systems thinking principles to model organisations and their decision-making process to improve the system’s functionality (Forrester, 1961).

While Forrester is considered the father of system dynamics, Austrian biologist Ludwig von Bertalanffy with his book “General systems theory” is often described as the father of systems thinking. He was one of the first scientists to recognise that elements within the human body form systems, which he defined as “a set of elements standing in interrelations among themselves and with their environment” (Von Bertalanffy, 1969, p. 252). What made this definition ground-breaking was that it emphasised the importance of interrelationships within a system, as he found that the different elements in a system could not be fully understood in isolation (Van Assche et al., 2019).

Russell Ackoff, an American management and complexity scientist, was also a system-thinking pioneer (Capaldi et al., 2019). Ackhoff defined a system as “a set of interrelated elements. Thus, a system is an entity which is composed of at least two elements and a relation that holds between each of its elements and at least one other element in the set. Each of a system’s elements is connected to every other element, directly or indirectly. Furthermore, no subset of elements is unrelated to any other subset” (Ackoff, 1971, p. 6). He explained that a system is not the sum of its parts, but rather the product of the interaction of the parts (Preuss, 2013). In later years he defined systems thinking as a way to examine “...relationships (rather than unrelated objects), connectedness, process (rather than structure), the whole (rather than just parts), the patterns (rather than contents) of a system, and context” (Ackoff, 2010, p. 6).

Another important contributor to systems thinking was the British manager, consultant and cybernetician Stafford Beer, whose integrating theory and practice pioneered the concept of management cybernetics, which is the application of cybernetic principles to the management of large organisations (Ramage & Shipp, 2020). Like the authors before him, he defined a system as a collection of interrelated elements that function to achieve a common goal. He emphasised the understanding of system behaviour, and the necessity to look beyond individual components and focus on the relationships and interactions among them (Beer, 1979).

However, Peter Senge, an American management scientist, popularised systems thinking in 1990 with his book "The fifth discipline: The art and practice of the learning organization". He saw systems thinking as a holistic approach to learning and organizational change (Senge, 1990). He defined systems as: "a perceived whole whose elements 'hang together' because they continually affect each other over time and operate toward a common purpose. The word 'system' descends from the Greek verb sunistánai, which originally meant: to cause to stand together" (Senge et al., 1994, p. 90). He defined systems thinking as "a way of thinking about, and a language for describing and understanding, the forces and interrelationships that shape the behaviour of systems" (Senge, 1990, p. 6-13).

A more recent contributor to systems thinking is American environmental scientist Donella Meadows, known for her work "Thinking in systems: A primer". She defines a system as "an interconnected set of elements that is coherently organized in a way that achieves something. ... a system must consist of three kinds of things: elements, interconnections and a function or purpose" (Meadows, 2008). It is a tool to "...help us manage, adapt, and see the wide range of choices we have before us. It is a way of thinking that gives us the freedom to identify root causes of problems and see new opportunities" (Meadows, 2008, p. 1-2).

Different geographers also applied and contributed to systems thinking in their research, as it offers a comprehensive and interdisciplinary approach to analyse the complex interactions between social and environmental systems (Mehren & Rempfler, 2022). This includes the interactions between elements such as land use (Bréthaut et al., 2019), climate (Ison & Straw, 2020), populations (Ramos-Quintana et al., 2019), water resources (Wan Rosely & Voulvoulis, 2022), and economic activity (Oliveira et al., 2020) within a specific region, which are studied to provide more effective solutions to challenges such as land use change, urbanisation, natural resource management, and climate change (Mehren & Rempfler, 2022).

Even though many other authors from various fields also use systems thinking, we will draw from the above-mentioned pioneers to establish what is defined as systems and systems thinking in this study. In this current study, systems refer to: elements which directly or indirectly interact with one another within a particular context to achieve a specific purpose. The assumption is therefore that the UNESCO MAB Biosphere Reserves form a system; the context of the study is the UNESCO AfriMAB network; and the elements within the system are the biosphere reserves and their related subsystems, which interact to create a network. The goal of this system is to achieve biodiversity conservation, sustainable development, and logistical support (Job et al., 2019).

Systems thinking is consequently regarded as an analytical tool that is used to understand the whole of the system, enabling the identification of opportunities and shortcomings, to ultimately predict and modify system behaviours and produce desired outcomes. This research will therefore adopt Behl & Ferreira's (2014) core principles of systems thinking, which include the holistic consideration of the entire system (i.e., all elements, components, and sub-systems), the understanding of its interconnections (i.e., having an insight into the relations and interdependencies between various system elements), and the importance of multiple perspectives (i.e., from a variety of expert's points of view).

While systems thinking is a useful tool to analyse complex systems, it is not perfect. For example, one of its benefits is also a shortcoming, namely its complexity, which makes understanding systems difficult (Capaldi et al., 2019). It is also a time-consuming approach, and its research usually takes longer than reductionist approaches (Salmon et al., 2022). In addition, it lacks standardisation as there is currently no standard methodology for conducting systems thinking research. As it is a relatively new field, there is also a limited understanding of its concepts (Chowdhury, 2023). As systems thinking takes a holistic approach, a large amount of data is required, and this can be difficult to obtain and interpret (Beasley, 2012; Salmon et al., 2022). These shortcomings can be overcome by firstly working with experts from various fields who bring a variety of expertise and perspectives; secondly, by incorporating feedback and adaptation into the research and data collection processes; and thirdly by combining systems thinking with other holistic research approaches and methods to help focus the study (Meadows, 2008). The social-ecological systems approach is an example of this, as it forms part of systems thinking but has more focused research methods.

3.3 Social-ecological systems

Social-ecological systems specifically refer to the complex adaptive systems created from the interaction between societal and ecological elements and processes (Farahbakhsh et al., 2022). These social elements include economy, politics, governance, and culture, while ecological elements frequently include climate, habitat, species, and biomes (Schlüter et al., 2019; Castillo et al., 2020). This research approach emerged in the 1970s when scientists became aware of the extent to which society and the environment are intertwined – how dependent humans are on ecosystems and how much ecosystems are influenced by human activities (Burns, 2008). This realisation occurred simultaneously in a variety of disciplines, including biology, sociology, ecology, anthropology, engineering, and geography (Colding & Barthel, 2019). As a result, social-ecological systems are also referred to as social-environmental systems, coupled human-natural systems, social-nature systems, human-environment systems, coupled human-biophysical systems, and human-ecological systems (Petrosillo et al., 2015; Virapongse et al., 2016; Schlüter et al., 2019). For this study, we will use the term social-ecological systems, as it is broadly accepted and used in seminal research frameworks (McGinnis & Ostrom, 2014).

The social-ecological systems approach is – relative to other approaches – fairly new. The term "social-ecological systems" was first coined in 1969 by American engineer Everett Dale Retzlaff, who used it (but did not specifically define it) in his dissertation titled "Applications of engineering systems analysis to the human social-ecological system" to describe societal and environmental crises brought on by the global over-consumption of natural resources (Ratzlaff, 1969). However, he did not publish further on the topic.

It was not until Canadian ecologist Crawford Stanley Holling started writing extensively on the core principles of social-ecological systems that the field started to truly emerge. Holling's work focused on the interplay between ecological and social elements to promote resilience and stability in the systems (Holling, 1973). One of his later contributions was the introduction of feedback loops: the outcomes of interactions within a system are assimilated into the system dynamics, resulting in new interactions and thus new outcomes (Gunderson & Holling, 2002). Holling produced many of the principles on which social-ecological systems are based, but while this was ground-breaking research in this field, he was not the first person to properly define the concept. The credit for this goes to Russian microbiologist B.L. Cherkasskii, who defined social-ecological systems as: "...two interacting subsystems: the biological (epidemiological ecosystem) and the social (social and economic conditions of life of the society) subsystems, where the biological subsystem plays the role of the governed object, and the social acts as the internal regulator of these interactions" (Cherkasskii, 1988, p. 321).

The field was popularised in 1998 by Fikret Berkes, a Canadian social-ecologist, and Carl Folke, a Swedish trans-disciplinary environmental scientist, with their book titled "Linking social and ecological systems" (Colding & Barthel, 2019). Here social systems refer to land and resource tenure systems, property rights, knowledge systems, and worldviews and ethics regarding the environment and natural resources. Ecological systems refer to the natural environment. Their statement that "the delineation between social and natural systems is artificial and arbitrary" set the tone for future research (Berkes et al., 1998, p. 4).

From this point forward there were many more influential social-ecological systems researchers, many of whom frequently co-authored peer-reviewed articles. One notable expert is Australian ecologist and sustainability scientist Brian Walker, who co-founded the Resilience Alliance, a global network of researchers, practitioners, and policymakers focused on advancing the understanding of resilience in social-ecological systems. In 2002 in a paper titled: "Resilience management in social-ecological systems: A working hypothesis for a participatory approach", Walker and his co-authors defined social-ecological systems as: "...complex, dynamic, and interconnected systems that encompass both human and natural components. They are shaped by multiple feedbacks and nonlinearities that emerge from interactions among social, economic, and biophysical components, and they exhibit path-dependent and threshold behaviour. Social-ecological systems are also characterized by uncertainty and surprise, as well as by the capacity of humans to manage and adapt to changing conditions. Social-ecological systems are thus not only subject to ecological and physical changes but also to social, institutional, and economic changes that can alter their resilience and adaptive capacity" (Walker et al., 2002, p. 558).

Lance Gunderson, an American ecologist, is another co-founder of the Resilience Alliance. Gunderson is a leading researcher in the sub-field of panarchy, which is a term used to describe complex social-ecological systems as having multiple levels of organization and interaction, ranging from individuals and organizations to ecosystems and the biosphere. These levels are connected through feedback loops, where changes at one level can trigger changes at another level, leading to adaptive change and development over time (Gunderson & Holling, 2002).

British geographer W. Neil Adger also made significant contributions to the understanding of the links between environmental change, human well-being, and social-ecological systems with the purpose of promoting resilience and sustainability. In his article "Social-ecological resilience to coastal disasters", he defined social-ecological systems as "a multilevel system comprising human and natural subsystems and the interactions between them, including the exchange of goods, services, information and people, and the feedbacks between human activities and the state and condition of the natural environment" (Adger et al., 2005, p. 578).

While there are many other contributors to the development and understanding of social-ecological systems, a particularly notable figure is American political economist Elinor Ostrom. Ostrom won the Nobel Memorial Prize in Economic Sciences in 2009 for her work on the governance of common-pool resources. Her research on the governance of natural resources has important implications for the social-ecological systems approach (Ostrom, 2009). According to Ostrom social-ecological systems are complex and interdependent systems that consist of both human and natural components, and are characterized by their ability to generate a wide range of goods and services that are essential for human well-being (Ostrom, 2010).

As is evident from the various definitions of social-ecological systems, the field has developed greatly since the 1970s and has been influenced by a large variety of disciplines. In this study we define social-ecological systems as a complex adaptive system whose natural and social elements interact to create specific outcomes, system characteristics, and feedback, resulting in a single multi-level, integrated system that changes over time.

While social-ecological systems is a valuable research approach it does share some of the disadvantages of systems thinking. It is still a highly complex research approach, and in places does suffer from a lack of research standardisation (Binder et al., 2013). Also, scientists from various disciplines studying social-ecological systems do not always share a single interdisciplinary vocabulary to communicate concepts used to construct and compare theories and explanations about the social-ecological systems (Castillo et al., 2020). This means that there is a limited amount of comparable knowledge in the field as the studies have different approaches, research methods, frameworks, and world views (McGinnis & Ostrom, 2014). Even though social-ecological systems as a research approach is narrower than systems thinking, it is still typically quite broad and difficult to analyse without a research framework to guide its analysis and orientate its outcomes (Binder et al., 2013).

3.4 Frameworks for analysing social-ecological systems

A research framework is a basic vocabulary of concepts, assumptions, values, techniques, and practices that allows a researcher to organise and understand research inquiries (McGinnis & Ostrom, 2014). Although several research frameworks are used in social-ecological systems research, according to Binder et al. (2013) only ten are properly conceptualised and general enough to be applied in a wide range of scientific scenarios:

- *Driver, Pressure, State, Impact, Response Framework*: it is used to describe the relationships between the origins and consequences of environmental problems (Troian et al., 2021). Elements of a system are classified as "drivers" that place "pressure" on the "state" of environmental or social systems and result in "impacts" that evoke certain "responses" in policy or individual behaviour (Quevedo et al., 2023). This framework has however been criticized for overly simplifying complex social-ecological systems (European Environment Agency, 1999).

- *Ecosystems Services Framework*: this framework aims to identify the various (often monetary) benefits that people derive from the natural environment (Sangha et al., 2018). This framework identifies, quantifies, and evaluates the different ways in which ecosystems contribute to society, such as the provision of clean water and food, the regulation of climate and diseases, and recreational and cultural values (Sangha et al., 2018). This framework is often used as a way to enlighten practitioners, managers and policymakers, about the benefits of protecting ecosystem services (Blahna et al., 2017). An important disadvantage of this framework is that it neglects the cultural bonds between society and nature (Sangha et al., 2018).
- *Earth System Analysis*: it is used to analyse complex interactions and feedback between the land surface, atmosphere, oceans biosphere, and cryosphere of the Earth, which when combined create a single whole system (Donges et al., 2021). This framework specifically focuses on the human impact on the Earth's systems (Brauch et al., 2011). Concerns have been raised about the framework practitioners' ability to fully understand each of the Earth's systems and the limits in the computing power to analyse them (Heymann & Dahan Dalmedico, 2019).
- *Human-Environment System Framework*: this is a heuristic tool used to structure the investigation of human-environmental interactions, processes, and dynamics. The overarching assumption of this framework is that there is a mutual dependence between two separate systems: social and ecological (Scholz & Binder, 2004). It emphasises different environmental feedback loops (e.g., perception, evaluation, and adaptation) impacting human systems (Le & Feitosa, 2012). One of the criticisms against this framework is that it has reductionist tendencies as it focuses on natural components but neglects social and cultural aspects (Blaikie, 2006; Folke et al., 2016; Eakin et al., 2017).
- *Material and Energy Flow Analysis*: it aids our understanding of how energy and materials flow through a system from the time it is extracted to its disposal or recycling (Kullmann et al., 2021). This analysis is an effective tool to identify systematic problems and develop sustainable systems. Its outcome is unusual and takes the form of schematic diagrams of energy and material movement throughout the system (Naohiro et al., 2015). A major disadvantage of this type of analysis is that it oversimplifies complex systems and has a rather narrow focus. It emphasises material flows and environmental

impacts but neglects the social or cultural dimensions (Pryshlakivsky & Searcy, 2021).

- *Management and Transition Framework*: this is an interdisciplinary framework originally designed for the analysis of water systems, management processes, and multi-level governance systems (Pütz & Job, 2016). It aims to improve the scientific understanding of system properties and to provide guidance for the implementation of transition processes that enhance adaptive capacity (Pahl-Wostl et al., 2010). This framework has not been applied widely (Pahl-Wostl et al., 2010; Binder et al., 2013).
- *Sustainable Livelihood Approach*: it facilitates the identification of practical priorities in terms of scope and objectives of development to enhance the lives of the vulnerable and poor (Morse & McNamara, 2013). This framework organises the elements that can restrict or enhance livelihoods and show how they are connected (Serrat, 2017). One of the weaknesses of this framework is that it does not consider ecological processes in the generation of livelihoods (Kimario et al., 2020)
- *Natural Step Framework*: this is a methodological tool used for strategic planning intended to move a company, organisation, or government towards sustainable development (Nattrass & Altomare, 1999). The framework is based on four principles that must be followed to achieve sustainability, relating to materials extracted from the Earth's surface; society's level of production; social and environmental degradation; and that society's needs are met (Everard, 2001). It is considered one of the best structures for thinking about the interaction between societal activities and nature, specifically within the context of sustainable development. The framework does not provide implementation strategies, or tools (Nattrass & Altomare, 1999).
- *Vulnerability Framework*: it aids the analysis of social-ecological systems where changes in the system can cause both social and natural elements to become vulnerable. The basic structure of this framework includes the connections between the social and environmental conditions and processes; the stress elements and obstacles that arise from these interactions; the areas where vulnerability resides; or where it is addressed within a system. These elements are interactive and scale-dependent (Turner et al., 2003). One of the major criticisms against this framework is the lack of a clear definition of what vulnerability encompasses (Auber et al., 2022).

- *Social-Ecological Systems Framework*: it was designed to help identify the driving elements and critical relationships within social-ecological systems. This framework is a multi-tier list of conceptual categories whose first tier includes resource units, resource systems, governance systems, and actors. These categories each have multiple layers, where lower-level tiers describe the characteristics of the higher-level tiers. Generally speaking, lower-level tiers can be added or removed depending on the research being conducted (Ostrom, 2009). As a result, this is one of the most versatile yet comprehensive frameworks for the analysis of interactions and outcomes in the study of social-ecological systems (Partelow, 2018). Previous research using this framework has been criticised for not considering how governance strategies impact societal inequalities. This however can be addressed by including it as a focus in the framework (Kimario et al., 2020).

Most of these frameworks have a unique approach to analysing social-ecological systems, as each originated and is used in various fields, including ecology, economics, resilience thinking, and geography (Hinkel et al., 2014). The aim of this study is to determine how the AfriMAB network can collectively position itself in its current institutional structures to better contribute to the CBD's 30X30 target and specifically improve its environment's resilience to climate change. An all-encompassing, well-defined framework that views social and ecological components as part of one system is therefore needed, one that also incorporates both the governance and social aspects. It is for these reasons that the social-ecological systems framework is considered to be the most appropriate for this study.

3.5 Social-ecological systems framework

The social-ecological systems framework originated from Elinor Ostrom's earlier Institutional Analysis and Development Framework. It was designed to describe how communities manage their natural resources (Voelzke, 2019). The goal of this framework was to prove that communities will solve social dilemmas that occur because of natural resource constraints (Ostrom, 2009). Ostrom and her colleagues (sustainability scientist John Anderies, econometrist Marco A. Janssen, and other members of the Resilience Alliance) collected empirical evidence in support of this hypothesis (McGinnis & Ostrom, 2014). Their results directly contradicted Garret Hardin's Tragedy of the Commons theory, which stated that when individuals have access to common natural resources, they act in their own interest and ultimately deplete a scarce resource (Hardin, 1968). Ostrom's findings on the other hand showed that communities develop self-organized institutions to manage such resources, also called collective action, without the need to privatise the resource. From this point onward the social-ecological systems framework was further developed, and in 2009 Elinor Ostrom won the Nobel Prize in Economics for her work on this topic (Ostrom, 2009).

Together, Ostrom and the Resilience Alliance tried to create a theory in the form of a list of social and ecological conditions that need to be fulfilled to encourage self-organising institutions. It quickly became evident that it is not possible to create such a one-size-fits-all list, as self-organisation within communities is dependent on complex interactions between social and ecological elements that differ from context to context. The social-ecological systems framework is therefore not a theory but can be used to explain causal relationships within a system (Schrogl et al., 2021). It also enables researchers around the globe to compare case studies across fields and disciplines, as they share a common conceptual vocabulary. The social-ecological systems framework is therefore considered to be a multi-tier list, or conceptual map, of potential key elements that can be used as a diagnostic tool to identify problematic system element interactions (Partelow, 2018).

The framework has seven first-tier categories: the social, economic, and political settings; related ecosystems; resource systems; resource units; actors; and governance systems (Figure 3.1). For example, within a social, economic, or political setting (e.g., UNESCO Biosphere Reserve Network) in a related ecosystem (e.g., African tropical forest), resource units (e.g., wood) are extracted from the resource system (e.g., biosphere reserve). The resource beneficiaries/actors (e.g., biosphere reserve board) manage the resource system through institutional structures and projects (e.g., climate change adaptation and mitigation projects). Many other second-tier elements (e.g., location and economic value) also have a role in the interaction (e.g., harvesting), and ultimately create certain outcomes (e.g., deforestation or forest conservation). During the framework's action situation, certain lessons are learnt and are fed back into the first-tier categories through feedback loops, to create the unique characteristics of a specific social-ecological system (McGinnis & Ostrom, 2014).

To customise the framework for this study, the third-tier categories were expanded by including documents relating specifically to biosphere reserves and climate change. These were:

- The 2020 "Statutory Framework of the World Network of Biosphere Reserves", which was created to guide biosphere reserves and clarify their function and purpose (UNESCO, 2020);
- The 10-year strategy document for biosphere reserves titled: "Lima Action Plan for UNESCO's Man and the Biosphere (MAB) Programme

and its World Network of Biosphere Reserves (2016-2025)", which contains a list of actions ensuring the effective implementation of the MAB strategy for this period (UNESCO, 2017);
- "The management manual for UNESCO Biosphere Reserves in Africa", was created in cooperation with several African biosphere reserve experts and refined during five workshops held with 110 biosphere reserve managers and experts. The purpose of this document was to elaborate on all the major components needed to make a biosphere reserve function in Africa (German National Unesco Commission, 2015);
- A document showcasing biosphere reserve activities and their contribution towards climate change mitigation and adaptation titled: "For life, for the future: Biosphere reserves and climate change" (Möller, 2011).

This study assumes that each element, in each category within the framework, contributes to the basic institutional structure of AfriMAB. The following sections elaborate on Ostrom's (2009) social-ecological systems framework as well as the incorporation of elements from the documents mentioned above (Figure 3.1). The sections are divided into external and internal influencing categories, followed by the interactions and outcomes sections of the social-ecological system framework. Some of Ostrom's (2009) second-tier categories are less relevant for this study and have therefore been excluded.

3.6 External influencing categories

The social, economic, and political settings and related ecosystems are influential categories that stand outside the focal system but still affect it. This category is used to describe external institutional contexts which affect the system. Ostrom's list of second-tier elements includes economic development, demographic trends, history, governance resource policies, market incentives, media organisations, and technology and science (Ostrom, 2009) (Figure 3.1).

Figure 3.1 Social-ecological systems framework with tiers adapted for this study.

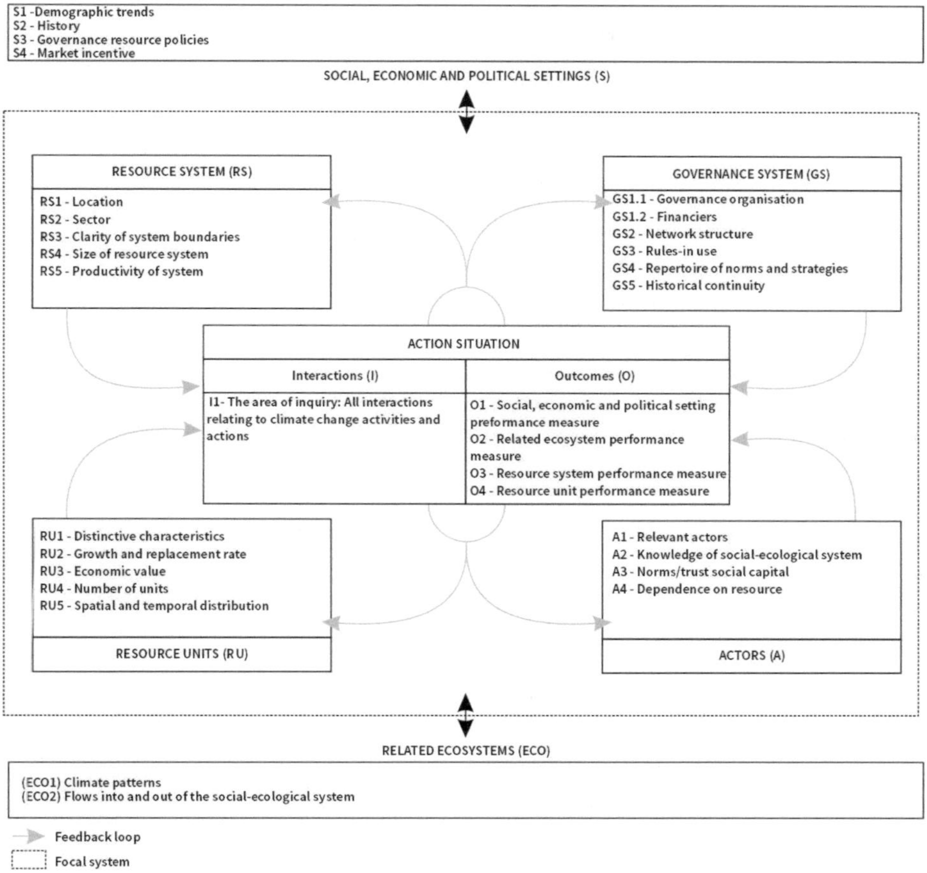

Source: adapted from McGinnis and Ostrom (2014)

3.6.1 Social, economic and political settings (S)

- *(S1) Demographic trends*: level of respondent involvement in AfriMAB.
- *(S2) History*: participation in past international events dealing with climate change declarations, targets, agreements, pacts, and action plans.
- *(S3) Governance resource policies*: attendance of recent international environmental conferences.

- *(S4) Market incentives*: impact of UNESCO MAB status to combat climate change.

3.6.2 Related ecosystems (ECO)

- *(ECO1) Climate patterns*: predicted impact of climate change in different zones of the biosphere reserve.
- *(ECO2) Flows into and out of the social-ecological system*: representation of different thematic networks per country.

3.7 Internal influencing categories

The focal system according to Ostrom (2009) contains the resource system, resource units, governance systems, and actors.

3.7.1 Resource system (RS)

- *(RS1) Location*: biosphere reserve's location.
- *(RS2) Sector*: number of biosphere reserves with climate change projects.
- *(RS3) Clarity of system boundaries*: this may be physical boundaries, but can also be the parameters of the issue being addressed within the biosphere (Ostrom, 2009). In this case, the parameters are created by the knowledge that exists about greenhouse gas emissions and the effect of climate change on the resource system. The third-tier category to be investigated is therefore:
  - Knowledge of CO_2 emissions emitted in biosphere reserve.
  - Knowledge of climate change impact.
- *(RS4) Size of resource system*: size of biosphere reserve; and area covered by climate change projects.
- *(RS5) Productivity of system*: year that the projects were first implemented.

3.7.2 Resource units (RU)

In this study resource units specifically refers to mitigation and adaptation projects produced within the biosphere reserve either by the biosphere reserve itself, or by partner organisations operating within the biosphere reserve.

- *(RU1) Distinctive characteristics*: type of projects and project categories.
- *(RU2) Growth and replacement rate*: initiation of climate change projects.
- *(RU3) Economic value*: the average cost of climate change projects.

- *(RU4) Number of units*: climate change projects number since 2000.
- *(RU5) Spatial and temporal distribution*: AfriMAB Biosphere Reserves with climate change projects, and the average duration of these projects.

3.7.3 Governance systems (GS)

The governance system is the internal institutional framework that governs the focal system. This includes rule-making organisation, network structure, rules-in-use, repertoire of norms and strategies, and historical continuity, which is a combination of Ostrom (2009) and the updated governance system categories in McGinnis & Ostrom (2014). The third-tier, and in some cases fourth-tier, categories in this section were mainly derived from Möller (2011).

- *(GS1.1) Governance organisation*: biosphere reserve role in climate change projects.
- *(GS1.2) Financiers*: this is not directly mentioned in the social-ecological systems framework (McGinnis & Ostrom, 2014), but is added here as financiers at various geographic levels often have a vested interest in the projects and how they are managed, and are therefore also a governance body (Möller, 2011). The third-tier category is:
    - Climate change project funders.
- *(GS2) Network structure*: the relationship between network levels.
- *(GS3) Rules-in-use*: the institutional rules (i.e. institutionalisation) that are implemented to address climate change within the AfriMAB Biosphere Reserve structure (McGinnis & Ostrom, 2014). The third-tier categories are:
    - Climate change in the management plan.
    - Existence of climate change strategies.
    - Climate change is the focus of the biosphere reserve.
    - Biosphere reserve has a budget for climate change.
- *(GS4) Repertoire of norms and strategies*: these are the rules and norms within the system that are implemented without necessarily being prescribed by the rule-making organisation (McGinnis & Ostrom, 2014). The third-tier categories are:
    - Climate change planning priorities in the participatory process.
    - Adapt, plan, implement, and evaluate processes of the management plan, and the frequency thereof.
    - Involve new stakeholders, including indigenous communities from outside the biosphere reserve.

  - Integrate policies of several government entities/departments/agencies.
  - Local climate change scenarios/predictions.
  - Long-term climate change monitoring.
  - New interdisciplinary research.
  - Generate new knowledge by combining results across projects.
  - Involve political/economic stakeholders in research design.
  - Innovative climate change education for children and students.
  - Create public awareness/understanding of climate change.
  - Mutual learning among stakeholders about climate change.
  - Sensitize decision-makers and donors to climate change.
- *(GS5) Historical continuity*: institutional survival during crises such as Covid-19.

3.7.4 Actors (A)

Actors (A) refers to any person or organisation involved in climate change mitigation or adaptation connected to the biosphere reserve.

- *(A1) Relevant actors*: individuals and organisations involved in climate change projects; stakeholders responsible for climate change projects; and stakeholders responsible for research.
- *(A2) Knowledge of social-ecological system*: knowledge of climate change scenarios; local communities' climate change awareness; and local companies' climate change awareness.
- *(A3) Norms/trust social capital*: strength of the relationship between biosphere reserve administration and locals; climate change project activity of non-biosphere reserve organisations; and future climate change project role of non-biosphere reserve organisations.
- *(A4) Dependence on resources*: livelihoods of the local community.

3.8 Interaction (I) and outcomes (O) of the action situation

This section is where the elements from all the categories and tiers interact to create the system as we know it. It has two categories which are: interactions and outcomes. The interactions section is this research's main area of inquiry, as it focusses on all the interactions relating to climate change within each AfriMAB Biosphere Reserve, especially the institutional structure which consists of all the sections of the social-ecological systems framework. In terms of outcomes, the following elements are investigated:

- (O1) Social, economic, and political setting performance measures.
- (O2) Related ecosystem performance measure.

- (O3) Resource system performance measures.
- (O4) Resource units' performance measures.
- (O5) Governance system performance measures.
- (O6) Actor performance measure.

As the social, economic, and political setting performance measures and related ecosystem performance measures are categories that are external to the focal system, their performance is measured using the self-fulfilling prophecy theory, and the Delphi research method. The rest of the elements are measured using respondents' self-rating of adaptive cycle characteristics.

3.9 Self-fulfilling prophecy theory and Delphi research method

The concept of self-fulfilling prophecy dates to ancient Greek Mythology: King Laius of Thebes was warned by an oracle that his son, Oedipus, would kill him. He consequently abandoned his son, and in the end, Oedipus was raised by another family, and eventually killed his father unknowingly (Laungani, 2007). A self-fulfilling prophecy is therefore a false prediction or belief that can be either positive or negative, that influences a person's behaviour in such a way that it becomes true (Stukas & Snyder, 2022). In 1948, self-fulfilling prophecy as an academic concept gained traction in social theory after Robert K. Merton's "The self-fulfilling prophecy" paper. Merton defined it as a false definition of a situation that evokes a new behaviour which makes the originally false conception come true (Merton, 1948). Self-fulfilling prophecy in academia is also known as the interpersonal expectancy effect, whereby a group or person's expectation of the behaviour of another person or group brings about the expected behaviour (Rosenthal, 2012). Researchers have proven this effect by studying the education system, the business sector, and even the judiciary system (Jussim, 2001). The assumption this study adopts is therefore that current beliefs about the future of a group of people, such as members of the AfriMAB network, may help to shape that future.

In this study, the Delphi method was employed to determine the preconceived ideas about climate change and climate change strategies within AfriMAB. The goal of the Delphi method is for a panel of experts to reach a group consensus on future developments from a variety of scenarios (Vogel et al., 2019). This is achieved through multiple rounds of questionnaires, analysis, and feedback to respondents. While this is generally a qualitative research technique, some answers can be quantified with the use of a Likert scale (Beiderbeck et al., 2021).

The Delphi research technique was developed in the 1950s by the Rand Corporation during an air force sponsored project that aimed to forecast technological capabilities for military and industrial purposes. The first Delphi project aimed to divulge "expert opinion to the selection, from the point of view of a Soviet strategic planner, of an optimal US industrial target system and to the estimation of the number of A-bombs required to reduce the munitions output by a prescribed amount" (Linstone & Turoff, 1975, p. 10). Its original goal was to help decision-making processes when the available information was lacking. The two creators of the method, Helmer and Dalkey, were never happy with the use of the term Delphi, as it does not convey the scientific rigour involved in the method (Adler & Ziglio, 1996). Despite their objections to the use of the term, it gained popularity after Theodore J. Gordon and Olaf Helmer-Hirschberg (1964) published the "Report on a long-range forecasting study", where the method was for the first time applied outside the military sphere (Gordon & Helmer, 1964; Linstone & Turoff, 1975). Since then, the method has been used in a variety of fields, including information systems, marketing, health, tourism, and spatial planning (Yoopetch et al., 2023).

One tool of the Delphi research technique is scenario planning, where experts are either used to postulate possible future scenarios, or pre-existing scenarios are presented to them, and they must reach a consensus on which scenario is most likely to happen. Scenario planning is a useful tool and has been advocated by scientists such as Gerhard Stiens, a German spatial planner (Stiens, 1977, 2003), and can for example be used to predict future forest development in a landscape (Sacher & Mayer, 2019).

The Delphi method has been refined and modified over time and has evolved into several methods that serve different purposes, e.g. the policy Delphi (also called the focus of decision Delphi), where the goal is to generate the strongest opposing viewpoints on a policy issue from a panel of experts, and the normative Delphi, which focuses on desirable goals and priorities instead of potential future outcomes (Yousuf, 2007; Meijering & Tobi, 2018). The basis of the Delphi method is to employ expert opinions, as it is believed that experts are more likely to be correct about trends in their field than non-experts. Experts can generally be defined as individuals who possess specialized knowledge and expertise in the subject being studied (Yousuf, 2007). In this study, we consider representatives from the AfriMAB network who possess knowledge and experience of the UNESCO Biosphere Reserve governance structure and their efforts to implement climate change mitigation and adaptation strategies, as experts.

Early during the development of the Delphi method, it was discovered that bringing experts together under one roof negatively impacts the results of the study. This is because the loudest expert often dominates the conversation, rather than the person with the most knowledge or the soundest arguments.

This was addressed by sending questionnaires to experts, which made anonymity, one of the method's four major components, possible. By sending them the questionnaires, they were able to freely express their opinions without fear of judgment from their peers (Rowe & Wright, 1999). This also allows a larger number of participants, and participation by members who are geographically dispersed (Yousuf, 2007). The second component of the Delphi method is iteration between questionnaires, as experts can now change their opinions if they feel that it is warranted (Rowe & Wright, 1999). The third component is controlled feedback: after each round of the questionnaire, the experts are provided with feedback from the anonymous group. This once again allows for all opinions to be heard, without the conversation being swayed by the opinions of the most vocal members. The fourth component is statistical aggregation: after multiple rounds of questionnaire iteration, the group consensus is determined by calculating the statistical average (mean/median) of the estimates provided by the panellists in the final round of polling.

Therefore, the final consensus can be viewed as a weighted average of the opinions of the respondent group (Rowe & Wright, 1999), regardless of whether the experts agree or disagree with one another. There is no threshold for what is considered high or low group consensus, as this depends on the research question and context – on some questions, there might be no consensus at all (Beiderbeck et al., 2021). In this study, high consensus is considered to be at least 60% agreement.

The disadvantage of using the Delphi method is that it takes time to convince experts to participate in the questionnaire, and then even more time between questionnaire rounds. The Delphi method has also been criticized for not creating more accurate results than any other method, and that consensus is reached through pressure from polarizing viewpoints (Gordon, 1994). Despite these shortcomings, this research method is still a powerful tool when used to answer appropriate questions about future developments, such as climate change (Yousuf, 2007). This study used the original Delphi method to measure the performance of the social, economic, and political setting, by predicting the impact of both COP15 and COP27 on the AfriMAB network (Figure 3.2).

The original Delphi technique was also used in the related ecosystem category to measure its performance, by asking respondents what they predicted the impact would be of a 2°C global temperature increase on each of the biosphere reserve

Figure 3.2 Incorporating the adaptive cycle and self-fulfilling prophecy theory in the theoretical framework.

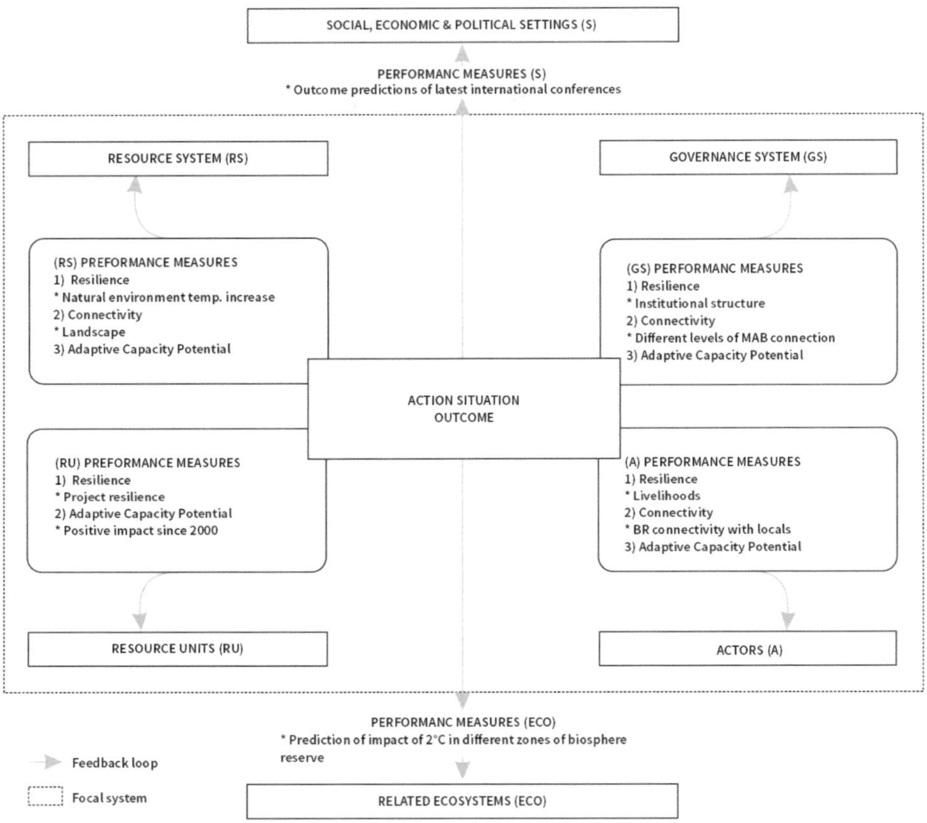

Source: own compilation

zones. They were also asked to predict the impact of a temperature increase on the different zones, and the likelihood of the three climate change scenarios as indicated by the IPCC. These scenarios were:

- *Climate change scenario: Business as usual.* If current greenhouse gas emissions continue, the global average temperature could increase to approximately 4°C from pre-industrial levels by 2100, surpassing the 2°C target by 2050. This could lead to extreme heat events, rising sea levels, declining biodiversity, polar ice melt, the destruction of coral reefs, fishery decrease, increasing poverty, societal health deterioration, and food shortages (UNEP, 2015).

- *Climate change scenario: Middling targets.* The Glasgow Pact aims to limit the global average temperature increase to 2°C by 2100, which would mean that by 2050 the global average temperature increase must remain below 2°C. This goal is supported by the UNESCO Biosphere Reserve Network and could be accomplished if climate change action is taken (UNFCCC, 2016).
- *Climate change scenario: Ideal targets.* The ideal aim of the Glasgow Pact would be to limit the global temperature increase to 1.5° C by 2100. The temperature increase trend would therefore need to plateau between 1° C and 1.5° C by 2050 (UNFCCC, 2016).

While the Delphi technique is useful to determine the performance and outcomes of external categories relating to a social-ecological system, a different approach is needed to determine the performance of internal categories. This study utilises the three characteristics used within the adaptive cycle and measure them through self-scoring questions in the research instrument.

3.10 Adaptive cycle

The Adaptive cycle is used to understand each focal system category's performance in the context of climate change. The adaptive cycle portrays the interactions of elements within a system and how they move. This can be used to predict when the system may change and when it may become stuck in difficult situations (Carpenter & Brock, 2008). The system is visualised as movement through a lazy-eight trajectory (Figure 3.3).

In Hollings's (1986) original model, the two scales through which it moves are connectivity (x-axis) and potential (y-axis). A system's potential is the range of options available that enables it to respond to changes in the system (Holling, 1986). Adaptive capacity shares similar functions to potential, except it is a better-defined concept with established indicators. This is the reason this study used the term "adaptive capacity potential" instead of "potential". Resilience, which was later added to the original model, is depicted on the z-axis. To be able to place a set of interacting elements on the adaptive cycle only two of the three categories are needed (Castell & Schrenk, 2020).

Figure 3.3 The adaptive cycle and its traps represent the interrelated relationship between adaptive capacity potential, connectedness, and resilience.

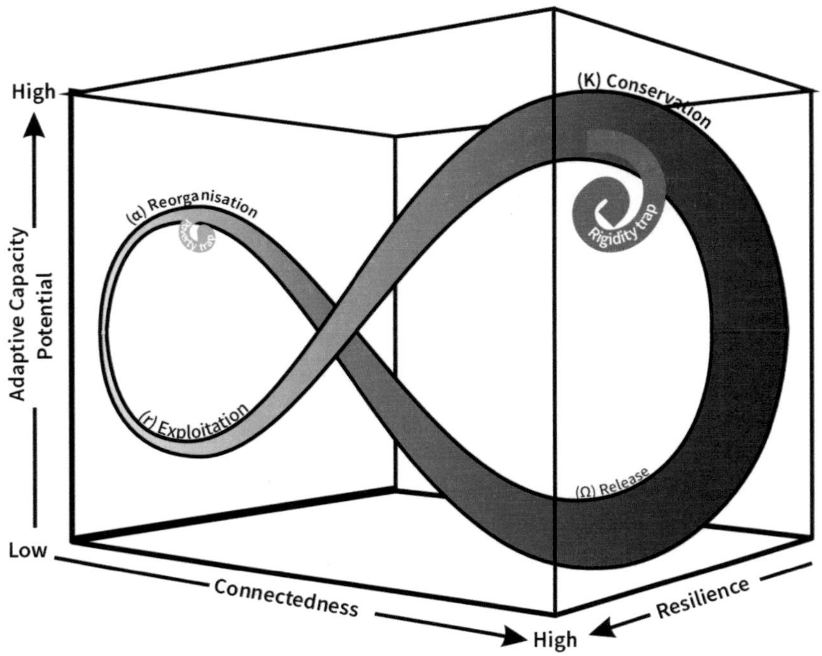

Source: Adapted from Carpenter and Brock (2008); Castell and Schrenk (2020)

The adaptive cycle describes the destruction and reorganization in a dynamic system. It has two main phases, divided into four stages (Figure 3.3). The first phase (front-loop) consists of the exploitation (r) and conservation stages (K). The goal of the r-stage is to grow and develop as quickly as possible (Holling, 1986). The system is open and draws from a large body and variety of ideas, inputs, innovations, learning opportunities, and connections (Fath et al., 2015).

The K-stage is where systems are relatively stable and where they remain the longest. Growth slows down as adaptive capacity potential and connectivity develop. The quality of inputs and connections becomes more important than the quantity (Fath et al., 2015). Here connectivity and adaptive capacity potential increases, but resilience decreases (Figure 3.3). This is because, over time, the systems become specialized, with dominant interactions making them rigid and increasing their vulnerability to disturbances (Holling, 1986). It is after this point that the system may enter a rigidity trap. These traps happen when institutions become highly connected, self-fortifying, and discouraging of diversity. They commonly occur where adaptive capacity potential is supposed to be high but because of rigidity not all opportunities are utilised, connectedness is high, and resilience is high (Carpenter & Brock, 2008). In the second phase (back-loop), after a disturbance completely disrupts the system, the system enters a release (Ω) stage where it is forced to utilise the built-up adaptive capacity potential (e.g., assets, information, and agency) to manage the event. The connectivity and adaptive capacity potential is high but decreasing, and resilience is low but increasing, as system governance finds new ways to manage the system to recover (Figure 3.3) (Castell & Schrenk, 2020).

As the system adapts and adjusts to the new norm it enters the reorganization (α) stage (Fath et al., 2015). The system draws on previous experiences to manage the disruption. During this stage connectivity decreases as old connections become redundant, and adaptive capacity potential and resilience increases (Castell & Schrenk, 2020). At this point there is a risk that the system may enter the poverty trap, where an institution becomes impoverished by circumstances beyond their control. The poverty trap is characterised by low connectivity and resilience, and where the adaptive capacity potential is not yet realised (Carpenter & Brock, 2008). Based on these interactions, only two of the three aspects are needed to determine the performance of a system.

3.10.1 Adaptive cycle performance measures

3.10.1.1 Resilience

The resilience component of the adaptive cycle emerged in the 1960s through ecological studies on predators and their prey, and ecosystem stability (Folke, 2016). The first scientist to adopt the term resilience in ecology was Holling (1973), who defined ecosystems as dynamic systems that are subject to constant disturbances. A system has key elements and interactions that drive it, yet keep it relatively stable even if these elements naturally wax and wane over time. However, when a disturbance is large enough, it will completely change the character of the system. Holling used the example of clear water lakes and how excessive nutrient inputs, fishing, and climate change can transform them into a turbid state, that is for example dominated by algae (Holling, 1973).

Since the term “resilience” was first introduced into ecology, two main schools of thought emerged as to what the term means. The first adopts an engineering

Figure 3.4 The difference between the way (a) ecological resilience and (b) engineering resilience is visualised after a disturbance.

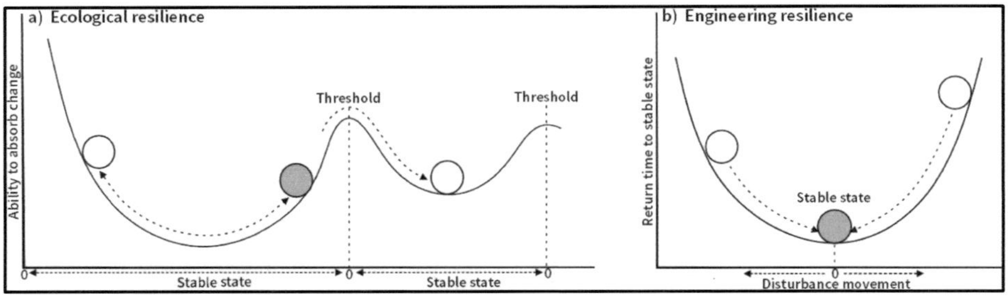

Source: Adapted from (Tri et al., 2017)

perspective of systems; that they are designed to have a single operating goal with one stable state (Figure 3.4). Here resilience is defined and measured in terms of the time it takes for the system to return to its supposed stable state (Holling, 1996). This definition, which is generally used in the fields of material engineering, physics, and control system design, was also employed by authors such as DeAngelis (1980) and Small et al. (2018).

Ecological resilience however is defined and measured in terms of the amount of disturbance a system can absorb while remaining in the same state. This definition assumes that there are multiple stable states that a system may enter (Holling, 1996). In this study we follow the example of authors such as Gunderson (2000) and Carpenter and Cottingham (1997) who use this approach in the fields of ecology and environmental management.

Both engineering resilience and ecological resilience can be visualised as a ball-in-cup heuristic, where the ball refers to the system and the cup refers to the stable state. Engineering resilience focuses on the slope of a single cup, measuring the time the ball needs to return to its centre (Figure 3.4). In ecological resilience there is more than one cup, and resilience is defined by the width of the cup – the amount of change that can be absorbed before the ball crosses a threshold and enters another cup. Here the system is considered to be dynamic and hence the "ball" is constantly moving within the "cup" (Walker & Salt, 2006) (Figure 3.4).

To influence a system's resilience, actors need to have adaptive capacity. Therefore, the adaptability of a system is a social component and actors' ability to adapt, self-organise, and predict changes determines whether a system can avoid moving into an undesirable state or recover from one (Walker et al., 2003). Adaptive capacity is often seen as a characteristic of resilience, but the term itself is intricate and needs to be considered on its own (Gallopín, 2006).

3.10.1.2 Adaptive Capacity Potential

Adaptive capacity has been studied in several fields, and therefore it also has many definitions (Jones et al., 2011). For example, McCarthy et al. (2001) and Nelson et al. (2007) define adaptive capacity as a trait of a system based on its resources enabling it to implement adaptation strategies (McCarthy et al., 2001; Nelson et al., 2007). Folke et al. (2010) describe it as an aspect of resilience, as a system can change and improve its coping capacity (Folke et al., 2010). Murthy et al. (2015) merely equates adaptive capacity to coping capacity. Coping capacity is the ability of a system to manage adverse conditions such as disasters and risks, with the resources they have available (UNESCO, 2010a). We will however adopt the IPCC's definition of adaptive capacity, as it not only views adaptive capacity in reaction to potential negative consequences but also considers potential opportunities. This definition is: "the ability of systems, institutions, humans, and other organisations to adjust to potential damage, to take advantage of opportunities or to respond to consequences" (IPCC 2014, p. 1758).

The determining elements of adaptive capacity have been widely researched. In a meta-study of adaptive capacity described in 276 academic articles that were published between 1980 and 2015, Siders (2019) identified 158 determining elements. However, most studies only utilise a small number of distinct, interrelated determinants in their adaptive capacity frameworks. These mostly include and relate to:

- *Assets:* access to, and control over, assets that enable the system to adapt more effectively to change (Jones et al., 2011). Assets can generally be defined as "items of value owned; in context, particularly those items that may be drawn on during times of need" (Siders, 2019, p. Supporting Information Appendix S3 3). Fair and equitable access to assets facilitate adaptive capacity as it allows a system to react more readily to change (Smit and Pilifosova, 2001). Assets can be divided into two categories: tangible assets, such as natural, physical and financial assets; and intangible assets, such as knowledge and skills (Jones et al., 2011). These definitions are understood differently in each framework, for example, Cinner et al. (2018, p. 117) focus on financial, technological, and service assets. Smit and Pilifosova (2001) however do not group assets together and instead distinguish between economic resources, technology, skills, and infrastructure (Keskitalo et al., 2011).

- *Flexibility*: the capacity of a system to incorporate and anticipate change into future planning and governance with new information or developments (Cinner et al., 2018; Jones et al., 2011). Flexibility is defined as the "capability of bending, yielding, or changing slightly (as opposed to transforming) under pressure; capacity for ready modification or alteration" (Siders, 2019, p. Supporting Information Appendix S3 17).
- *Information and knowledge*: information is defined as "knowledge communicated concerning a fact, subject, event" (Siders, 2019, p. Supporting Information Appendix S3 20). The adaptive capacity of a system depends on access to information and understanding the problem (Jones et al., 2011). The availability of information on past and future risk scenarios enables the creation of appropriate adaptation and mitigation strategies enhancing the adaptive capacity of the system (Brooks et al., 2004). However, obtaining information is not enough; it has to be converted into knowledge, and knowledge into action (Smit and Pilifosova, 2001).
- *Learning*: is closely related to the information and knowledge determinants, as it relates to the actor's ability to create, absorb, and process new information about a problem, and ways to mitigate and adapt to uncertainty (Cinner et al., 2018).
- *Innovation*: is to "create or bring in something for the first time, to change a thing into something new" (Siders, 2019, p. Supporting Information Appendix S3 21). It relates to actors' ability to change existing practices, behaviours and resource usage and even adopt new ones if necessary. This relates to Schumpeter's (1942) creative destruction theory that states that established processes are constantly destroyed in favour of new and more efficient ones; this is called creative destruction (Schumpeter, 2013). Innovations at both macro/high tech (national) and micro/low tech (local) scales are vital to keep the system dynamic and functioning (Jones et al., 2011).
- *Agency*: refers to the ability of actors to respond to problems individually or collectively, as they see fit. An organisation with agency has the power and freedom to mobilize other determinants of adaptive capacity without any constraints (Cinner et al., 2018).
- *Social organisation or institutions*: a social organisation is the way society is organised to enable cooperation, collective action, and knowledge sharing (Jones et al., 2011). Formal and informal

relationships between organisations, communities, and individuals can enhance or inhibit adaptive capacity. Institutions provide the societal rules that govern behavioural and organisational structures (Ostrom, 2005).

The determinants that will enhance a system's adaptive capacity depend on the area covered, the scale, and its structural characteristics. Appropriate indicators for assessing adaptive capacity must therefore be identified; this can be accomplished by answering the following four research questions (Brooks et al., 2004):

- What elements determine whether actions are taken to reduce vulnerability?
- Can we assess these elements to measure the capacity of a system to implement these actions?
- What are the internal and external obstacles to implementing these measures?
- How can capacity constraints be removed to enhance adaptation?

From the elements mentioned above it becomes clear that another major consideration for the performance of a system is connectivity.

3.10.1.3 Connectivity

Connectivity in resilient systems became a prominent research topic in recent years. In general, connectivity can be defined as the degree to which different elements in a system are either connected or independent. Within social-ecological systems this has mainly been studied in the field of system dynamics, which is an exciting new research frontier (Cinner & Barnes, 2019).

The role of connectivity within a system is to improve access to critical resources and services, as well as facilitate mobility and flexibility in response to disruptions. However, an overly connected system can harm the system, as it may allow it to become rigid and inflexible, preventing learning and innovation (Biggs et al., 2015). In an environmental context, connectivity within a landscape is a vital feature. It refers to the degree of connection between the various natural patches of land in the landscape matrix through natural corridors (Newsome et al., 2013). A low degree of connectivity in the landscape due to landscape transformation (from natural vegetation to urban, agricultural, or industrial areas) has negative environmental consequences, such as the loss of habitat necessary for biodiversity, the obstruction of migration patterns, the impediment of climate change related movement because of roads and fences, and the sedimentation of waterways (Taylor et al., 2006; Job et al., 2022). High connectivity helps ecosystems to remain healthy by aiding species migration and supporting the continuation of key ecological processes (Stoll-Kleemann & Welp, 2010). However, high connectivity in the landscape can also be negative as it allows disturbances such as fire, disease, and pollution to spread more readily than in less connected landscapes (Biggs et al., 2015).

Based on this, the following questions need to be answered in this study:

- To which organisations is AfriMAB connected?
- How strong are the connections between the different UNESCO Biosphere Reserve levels?
- How strong are the connections between the different AfriMAB Biosphere Reserves and local communities and organisations?
- How connected is the landscape within the AfriMAB Biosphere Reserves?
- Are the different biosphere reserve zones functioning as mandated by the ICC?

In both social and ecological networks, at some point an increase in connectivity starts to negatively impact resilience (Stoll-Kleemann & Welp, 2010). This relationship between resilience and connectivity can be seen in Holling's (1986) Adaptive Cycle. However, while we acknowledge this, in this study, we regard connectivity as positive, unless proven otherwise.

The question that therefore arises is: What key performance measures should be present in a biosphere reserve network as a social-ecological system, to enable it to collectively position itself, within its current institutional structures, to improve its contribution to the CBD's 30X30 target, and specifically improve its environment's resilience to climate change? The Resilience Alliance (Biggs et al., 2015) offers the following principles:

- *Maintain diversity and manage redundancy*. Redundancy is when a system has many different components, such as in the r-stage of the adaptive cycle. The presence of multiple components that have a similar function enables it to compensate when some components fail or are lost, increasing the system's overall resilience. For example, it is better to have a variety of projects relating to climate change rather than only one. It is also important to have a variety of governance systems, such as independent organisations, working with the biosphere reserve, as it provides diverse sources of knowledge (Kotschy et al., 2015).

- *Manage connectivity.* A connected system can be both positive and negative, and therefore it is important to manage it. This is done by analysing both its social and landscape connectivity to identify important interactions, so that degraded connections can be restored, and existing connections optimised (Dakos et al., 2015).
- *Manage slow variables and feedback.* There are many ways in which different system elements can be connected and interact with one another to create different outcomes. The first step is to identify slow elements and feedback loops, and then to strengthen these interactions to remain in the desired "cup", monitor these slow variables, and finally establish governance structures that respond to the information (Biggs et al., 2015).
- *Foster complex adaptive systems thinking.* Admit that a system is complex and that it is part of a web of interacting elements that can be managed to achieve a desired system state (Turner & Baker, 2019).
- *Encourage learning.* Social-ecological systems are constantly changing and adapting, and to effectively address undesirable system states, it is important to constantly expand the knowledge base. This is achieved by valuing diverse sources of knowledge when developing solutions (Fazey et al., 2020).
- *Broaden participation.* The engagement of all relevant stakeholders is fundamental to resilience as it leads to trust, participation, and broader support (Wamsler, 2017).
- *Promote polycentric governance.* When multiple governance bodies interact to achieve a goal within a specific location, the collaboration between the different institutions improves connectivity and learning. It also helps to deal with a disturbance quickly and effectively (Amaruzaman et al., 2022).

The presence of these elements in our adapted social-ecological systems framework will therefore be investigated to determine what climate change adaptation or mitigation strategies are being adopted in the AfriMAB network, and what institutional changes can be made to better incorporate them.

3.11 Concluding remarks

To answer the research question, we needed to firstly determine the best possible way to analyse a complex system, using systems thinking approaches, and it was found that an adapted version of the social-ecological system offers the best solution. Secondly, we had to establish the elements which would be investigated, forming the institutional structure of the AfriMAB Biosphere Reserves. These elements were identified using key documents guiding biosphere reserves, which we attached to the social-ecological systems framework. The details of how the data was collected to further populate the framework will be discussed next in the methodology and methods chapter.

4 Methodology and methods

4.1 Preliminary remarks

In this chapter we elaborate on the methods and methodology used during this study to determine how the AfriMAB Biosphere Reserves can collectively position themselves within their current institutional structure to better contribute to the CBD's 30X30 target and specifically improve their environment's resilience to climate change. The objective of this is to ensure that this research is reliable, valid, and replicable so that it can contribute to the academic body of knowledge on climate change actions within biosphere reserves. In this chapter we will first state the philosophical research assumptions on which this study is based. This is followed by the research paradigm, research design and strategy, literature review, data collection instruments, sampling strategy, data analysis methods, and concluding remarks.

4.2 Philosophical research assumptions

Philosophical research assumptions are the fundamental beliefs, values, and perspectives that underlie a research study. These assumptions shape the way researchers approach their queries and influence the methods of analysis that they use, which include includes ontological, and epistemological considerations (Garrick, 1999).

This study has an empirical research approach, as we rely on data collected directly from the real world, rather than knowledge gained either intuitively, through authorities, or through logical reasoning (Flynn et al., 1990; Bires & Raj, 2019; Pereira, 2023). Empiricism refers to knowledge gained through lived experiences and sensory learning (Bryman, 2016) and implies that real-world observations were made through a planned and systematic process, aiming to acquire specific information. This includes decisions about what to observe, whom to observe, how to observe, when to observe, and finally, how to analyse the data collected, and how to interpret it (Patten & Galvan, 2019). Empirical studies are rooted in the use of direct first-hand experiences as the basis for understanding and knowledge (Simon, 2017). The goal of empirical research is to generate knowledge by exploring, describing, or explaining behaviours observed and collected from the real world. This leads to a more systematic and scientific understanding of phenomena (James Madison University, 2023). Examples of empirical studies on African biosphere reserves include: "Potential and challenges of the Serengeti-Ngorongoro Biosphere Reserve, Tanzania" by Botha et al. (2021); "Local ecological knowledge on climate change and ecosystem-based adaptation strategies promote resilience in the middle Zambezi Biosphere Reserve, Zimbabwe" by Kupika et al. (2019b); and

"Governance of the Pendjari Biosphere Reserve in view of climate variability" by Gbénahou and Sokemawu (2020).

The ontological assumption of this study establishes how researchers regard the nature of social entities. There are two major philosophies: a social entity can be seen as an object with a reality external to social actors, which is called objectivism or, they can be seen as social constructs created by the actions and perceptions of the social actors, which is called constructivism (Bryman, 2016). This research is based on constructivism, as we accept that the social actors within the AfriMAB network have constructed the reality in which they find themselves through their participation and actions in the network.

The epistemological considerations used within this study relate to the methods of gaining knowledge and the type of knowledge gained (Fernández-Fernández, 2021). There are generally two major epistemological doctrines[2]. The first is positivism, which is often used in natural sciences to study a variety of natural and social topics. This knowledge aims to generate a hypothesis that can be tested and used to explain and predict interactions (Crossan, 2003). Such research is often context-free, and the researcher is independent of the subject being studied. It is therefore assumed that the research is value-free and unbiased (Clark, 1998). The nature of reality is considered to be single, objective, and tangible, and the research aims to obtain facts that can be used to form the basis of general laws (Bryman, 2016). The second major doctrine is interpretivism, which is often used in the social sciences. In this case the purpose of acquiring knowledge is to gain an in-depth understanding of interactions and real-world situations. Here the nature of reality is socially constructed, and as multiple participants are included, it is accepted that each has their own reality that must be analysed. Research is context-dependent, and the researcher interacts with those being researched. The researchers therefore acknowledge that the research is value-laden, and biases could be present (Schwartz-Shea & Yanow, 2012; Introna et al., 2016).

This research will adopt an interpretivism doctrine, as the overall aim is to understand the institutional structure(s) of biosphere reserves in the AfriMAB network, based on knowledge gained by interacting with its members. The advantage of using this doctrine is that the research includes the diverse views of participants with differing social, political, ecological, and economic contexts, which allows for a more in-depth understanding of the structure and potential improvements of biosphere reserves (Bryman, 2016). Consequently, we can better understand participants' thoughts, perspectives and opinions, which provide insight during the analysis phase (Pham, 2018). Interpretivism does however have disadvantages, for

[2] The definition of a doctrine is: a set of belief held and taught by researchers and philosophers (Oxford English, 2023).

example that one cannot generalise the results and apply it to other people and contexts. The subjective nature of interpretivism can also be a problem; research outcomes are often influenced by the researcher's own ways of thinking and cultural experience (Pham, 2018). Despite these shortcomings, the advantages outweigh the disadvantages and make it the more appropriate epistemological approach for this study. Two studies that applied both ontological constructivism and epistemological interpretivism are: "Theory and practice of conservancies: Evidence from wildlife management areas in Tanzania" by Kimario et al. (2020); and "Exploring challenges for the sustainable management of wildlife resources in Matusadonha National Park and its hinterland" by Kudakwashe et al. (2023).

4.3 Research paradigm

A research paradigm is the standard model or methods of conducting research within a discipline (Blaikie & Priest, 2017). The two main research paradigms are the qualitative and quantitative approaches. The qualitative research method is mainly used to investigate narrative related information, and is further characterised by the following (Hennink et al., 2010):

- *Aim*: The goal of qualitative research is to generate hypotheses, theories, or deep insights into social phenomena with an inductive research approach. It emphasises acquiring an in-depth understanding of human behaviour within their context (Given, 2008).
- *Focus*: Qualitative research explores complex phenomena, often in a natural setting, by examining subjective experiences, attitudes, and behaviours of individuals or groups (Hennink et al., 2010).
- *Data type*: Qualitative data is non-numerical: it is typically textual, visual, or auditory in nature. Examples include interviews, literature reviews, and open-ended survey responses (Taylor et al., 2015).
- *Sampling*: Qualitative data often uses purposeful or snowball sampling, focusing on participants who can provide rich, in-depth information about the research topic (Maruster, 2013).
- *Data analysis*: Qualitative data analysis is interpretative and involves identifying themes, patterns and meaning within the data. Techniques such as thematic and narrative analysis, as well as coding are commonly used (Hennink et al., 2010).

The quantitative research approach is followed to analyse quantifiable information, and is characterised by the following:

- *Aim*: the aim of quantitative research is to generalise findings from a sample to a larger population providing numerical insights and proving a hypothesis or theory. This is the deductive research approach.

It emphasises replication, measurement, and statistical analysis to draw conclusions (Balnaves & Caputi, 2001).

- *Focus*: quantitative research seeks to quantify relationships, test hypothesis, and identify patterns through the analysis of numerical data (Oflazoglu, 2017).
- *Data type*: quantitative data can be analysed using statistical methods. Examples include the Likert scale, numerical measurements, and test scores (Harkiolakis, 2017).
- *Sampling*: quantitative research often uses random stratified sampling techniques to ensure that the sample is representative of the larger population (Bryman, 2016).
- *Data analysis*: quantitative data analysis involves statistical techniques to analyse correlations, patterns, and causality. Common techniques include hypothesis testing and descriptive statistics (Balnaves & Caputi, 2001).

This study will adopt a mixed methods research approach, where qualitative and quantitative research approaches are utilised within the same study. This approach is characterised by the following:

- *Aim*: the aim of this approach is to provide a comprehensive understanding of a research problem by integrating qualitative depth with quantitative breadth (Schoonenboom & Johnson, 2017).
- *Focus*: it integrates the qualitative exploration with quantitative evidence which provides a holistic perspective of the research problem (Bryman, 2016).
- *Data type*: both qualitative and quantitative data are collected. There are several ways these aspects can be combined to accomplish this (Tashakkori & Teddlie, 1998).
- *Sampling*: both purposeful sampling for the qualitative part and random sampling for the quantitative part can be done, ensuring both depth and representativeness (Tashakkori et al., 2020).
- *Data analysis*: the combination of qualitative and quantitative data allows complementarity data collection with the aim to elaborate, enhance, illustrate and clarify results from the one method by using results from the other (Schoonenboom & Johnson, 2017).

According to Morse (2016) usually either the qualitative or the quantitative approach is dominant within a mixed methods research enquiry (the dominant method indicated by either QUAL or QUAN). Where QUAL is dominant, the quantitative (quan) approaches are supplemental, and where QUAN is dominant the qualitative (qual) approaches are supplemental. Johnson and Christensen (2019) states that it is possible for the QUAL and QUAN research approaches to have an equal weight in the study, but this is rarely the case. Qualitative and quantitative research can then either take place simultaneously (+) or sequentially (→) (Schoonenboom & Johnson, 2017). Overall, each mixed method aspect mentioned above has a different balance of quantitative and qualitative elements.

Qualitative/quantitative elements can be integrated in three areas: the data type, sampling, and data analyses (Bryman, 2016). In this study each aspect has its own qualitative/quantitative integration formula:

- *Data type*: when considering the type of data collected, the majority was quantitative in nature and could be expressed statistically, on a Likert scale, or as test scores. This was simultaneously supplemented with qualitative inquiries from an open-ended survey question (Morse, 2010). The qualitative/quantitative relationship can therefore be expressed as: QUAN + qual.
- *Sampling*: the individuals who were asked to complete this questionnaire were selected based on their knowledge of climate change projects and their biosphere reserves. The few participants who did not fulfil these criteria or did not have this knowledge identified suitable individuals to replace them (snowball sampling). Therefore, overall, the sampling of respondents was more qualitative in nature. However, the initial sample population was obtained from an email list and respondents were invited to participate (voluntary response sampling), therefore there was also a simultaneous element of random sampling, which is a characteristic of the quantitative research approach (Bryman, 2016). The qualitative/quantitative relationship can therefore be stated as: QUAL + Quan.
- *Data analysis*: the collected data was analysed using mostly quantitative statistical calculations such as descriptive statistics. This was simultaneously supplemented with a thematic analysis of the qualitative data. The qualitative/quantitative relationship can therefore be described as: QUAN + qual.

Examples of other studies that use this approach to investigate biosphere reserves in Africa are: "Green economy initiatives in the face of climate change: experiences from the Middle Zambezi Biosphere Reserve, Zimbabwe" by Kupika et al. (2019a), and "Evidence and impact of climate change on South African national parks. Potential implications for tourism in the Kruger National Park" by Dube and Nhamo (2020).

4.4 Research design

A research design provides a framework for the collection and analysis of data and relates to the causal relationships between elements, enabling the understanding of behaviour and meaning as well as a temporal appreciation of social phenomena and interactions. The five most prominent research designs are: experimental, cross-sectional, longitude, case study, and comparative designs (Bryman, 2016). An experimental research design aims to maximize precision and draws specific conclusions to prove or disprove a hypothesis by conducting research objectively and in a controlled manner (Bell, 2009). A cross-sectional research design analyses data from a population (more than one case) at a single point in time and compares different patterns of variable interaction (Wang & Cheng, 2020). A longitudinal research design is a repeated cross-sectional study where participants are largely or entirely different for each sampling session (Menard, 2007). A case study design deals with a single case. A case relates to a single location, event, or phenomenon that is explored within its natural context, to provide an in-depth understanding of a real life issue (Crowe et al., 2011). A comparative design study compares two different cases (Mello et al., 2021).

This study adopted a cross-sectional research design with a longitudinal element as it allows the inclusion of multiple cases; in this study each case is a biosphere reserve located in the AfriMAB network. Two questionnaires, each containing a different set of questions, were distributed to the same population at two consecutive occasions. The purpose of this was to follow up, clarify or extend the information collected in the first questionnaire. Some questions were repeated to determine whether respondents' answers had changed over time, signifying a longitudinal element. The cross-sectional design is based on quantifiable data, with the aim to establish patterns of association between cases and their variables (Wang & Cheng, 2020).

Research results were then structured in tables, where cases (biosphere reserves) were related to the variables (data elements) that were collected. This is how most of the data collected during the survey period was conveyed. The purely cross-sectional research design has some disadvantages, for example the fact that it only measures one point in time and therefore does not measure situations as they develop over time (Carlson & Morrison, 2009). However, by sending a second questionnaire to the same population, these drawbacks are reduced. Other authors working on biosphere reserves in Africa also use the cross-sectional research design, such as: "Determinants of household dietary diversity in the Yayo Biosphere

Reserve of Ethiopia: An empirical analysis using sustainable livelihood framework" by Huluka and Wondimagegnhu (2019), and: "Differences by age group in gender-related attitudes in Oromia, Ethiopia: Findings from a cross-sectional study of women of reproductive age" by Millar et al. (2023).

The research design followed during this study is stated below. It divides the study into four phases and provides a broad schedule (Figure 4.2).

- *Phase 1, Year 1*: the development of study's project proposal, including the problem statement and aim and objectives; submission to the university.
- *Phase 1, Year 2*: the initial literature review commences that would continue throughout the research period.
- *Phase 2, Year 2*: the expert review of the questionnaire that was created during the previous phase takes place. A pilot study is then conducted, after which the research instrument is improved. The AfriMAB email list is obtained, and the questionnaire is distributed online to the network. The data is collected, processed, and written in a research report.
- *Phase 3, Year 2/3*: based on the findings in the report of the first questionnaire round and further review of the literature, a second questionnaire is created. This is subjected to a quick expert review and pre-test before it is re-distributed to the same respondents.
- *Phase 4, Year 3*: the results are processed, and the dissertation is written, reviewed, and submitted.

Figure 4.1 Phases of the research design used in this study.

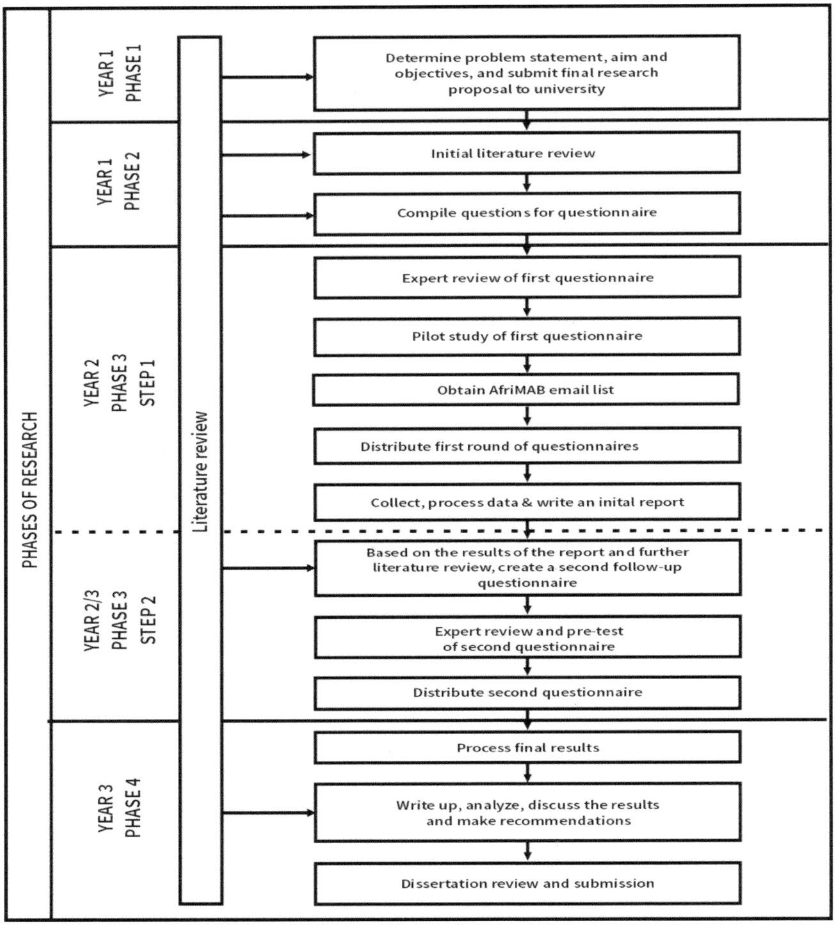

Source: Own compilation

4.5 Literature review approach

A literature review is an overview of published research and helps the researcher place their work in context with other published work and gain a better understanding of what they are studying (Zamanzadeh et al., 2015). There are two main forms of literature reviews, namely systematic and narrative. A systematic literature review is a replicable, scientific, and transparent process that leaves an auditor's trail for other researchers to follow. It is systematic and is highly focused (Bryman, 2016). The narrative review, which is generally broader, is the more traditional approach. The purpose of the narrative review is to gain understanding off a topic and enrich and support the arguments within the study (Demiris et al., 2019).

The narrative research approach is used in this study because its objective is to better understand the key elements and their interactions. However, we also incorporated elements of the systematic review process. Three questions needed to be answered through the literature review: First, how global events relating to climate change, nature conservation, and sustainable development have influenced biosphere reserves' creation, evolution, and current governance structures. Second, which theoretical framework is the most suitable to analyse this study's complex systems. Third, the literature review needed to clearly state the elements of the institutional structures within the framework to be investigated.

To answer these literature-based questions, online search engines such as Google Scholar, Scopus, and ResearchGate were used. The keywords used in the literature search in chapter two included: climate, climate measuring history, climate change, climate change research, nature conservation, conservation history, international environment movement, international environmental events, biosphere reserve history, biosphere reserve documents, and biosphere reserve management. In chapter three, the literature review keywords included terms such as: systems thinking, social-ecological systems, social-ecological systems frameworks, Delphi study, self-fulfilling prophesy, adaptive cycle, adaptive capacity, resilience, and connectivity. The main forms of literature that was searched were published research articles, books, book chapters, official organisational documents, and academic theses/dissertations.

This study aimed to use as many primary sources as possible. Where this was not possible, published works with high (20<) citation scores were preferred, or works with a more recent publication date. Published works that were excluded from the review that met these criteria, either did not contain relevant information, or repeated information already known. While we aimed to be exhaustive with the review, it was not possible to cite every published article on each of the many sub-topics. The reviewed literature was ultimately used to create the data collection instruments that will be discussed next.

4.6 Data collection instruments

Data collection instruments are tools or techniques used by researchers to gather data from participants in a systematic and organized manner (Kabir, 2016). There are several approaches to designing data collection instruments, each suitable for different research questions and methodologies. These include surveys (closed or open-ended questions), interviews (structured, semi-structured or unstructured), observation (participant or non-participant observation), focus groups, document analysis, and experiments (Bryman, 2016). While each of these instruments have its merits, this study will adopt the survey research approach.

The survey approach was chosen because of the nature of the research enquiry. It makes it possible to simultaneously reach participants scattered across the African continent. This approach was also an attractive research instrument because when this project was conceived, the Covid-19 pandemic was still in full effect, and it was still uncertain how it would develop in the future. The survey approach reduced physical contact with respondents and did not require traveling. Further advantages of the survey approach are that it gives respondents anonymity, allowing them to speak freely. The disadvantage of using surveys is that the answers are usually short and the response rate is usually low (Wood & Kerr, 2006).

The survey type chosen for this study was the questionnaire, and more specifically, an online questionnaire. This type of questionnaire is presented in a digital format that is accessible through a survey platform, in this case soscisurvey.de. Participants were invited to complete the online questionnaire through an invitation sent by email. An advantage of online questionnaires is that they are convenient for both the researcher and the respondent. The questionnaire can be distributed over a vast geographic area, and to many respondents. Furthermore, online questionnaires have automated data collection and organisation that simplifies the analysis process. The online questionnaire was considered an appropriate tool considering the uncertainties with Covid-19 pandemic, but it would have been the preferred instrument even without the pandemic due to the geographical distribution of the participants. An African biosphere reserve study that also used online questionnaires to gather information is: "Developing sustainability indicators for the Kogelberg and Cape West Coast Biosphere Reserves, South Africa" by Tucker (2013).

For this, we distributed two different questionnaires, six months apart, to the same research sample; it included an element of the Delphi research technique where "experts" were asked to predict the likelihood of future events. Some questions were asked in both questionnaires to determine whether the participants would reach a consensus (Taghipoorreyneh, 2023).

All the research measures/questions in the surveys were based on information gained from the literature review and theoretical framework. The exact measures that the questionnaires had to include are explained in chapter 3. The questionnaires both contained open-, semi-open and close-ended questions. Open-ended questions allow respondents to answer a question any way and length they want (Edwards, 1997). The advantage of open-ended questions is that they provide more thorough and diverse answers, and the disadvantage is that coding these answers usually takes longer (Wood & Kerr, 2006). Semi-open-ended questions are questions that ask something specific but provide an opportunity to expand on that specific element. These questions have the advantage of focusing a respondent's answer, however that means the respondent does not have the opportunity to introduce new concepts (Glerum et al., 2014). Close-ended questions are questions that provide several possible answers and the responded must pick one. They are easily coded but lack the data richness of the open-ended questions (Edwards, 1997). Each questionnaire was structured as follows:

4.6.1 Questionnaire 1:

This questionnaire contains 33 questions and consists of 9 (27% of the questionnaire) open, 3 (9%) semi-open, and 21 (64%) close-ended questions (Appendix A, p. 195). Each page of the questionnaire covers a specific research topic.

- *Page 1*: abstract, and permission to contact participant for the second questionnaire.
- *Page 2*: four questions about the respondent's level of involvement in the biosphere reserve network, and details about their country, biosphere reserve, and biosphere reserve's network affiliation.
- *Page 3*: provides the first business-as-usual climate scenario and asks fifteen questions regarding the biosphere reserve's current and past climate change adaptation and mitigation institutionalisations, strategies, and projects.
- *Page 4*: establishes the 2°C climate change scenario and asks seven questions about the probability of this scenario happening, and what organisations outside the biosphere reserve's management are contributing. This scenario requires all internal and external biosphere reserve stakeholders to work together, and it is therefore important to establish the external organisation's contribution.
- *Page 5*: contains the last 1.5°C climate change scenario, and four questions about the probability of this scenario and the contribution of

the respondent's organisation, as well as ways to increase the number of climate change projects and other ways they can help to limit the global temperature increase.

- *Page 6*: three questions about future scenarios they believed are most likely, and placing their organisation where their efforts to address climate change fit.
- *Page 7*: thank participants for sharing their knowledge and expertise.

4.6.2 Questionnaire 2:

This questionnaire contains 36 questions and includes 10 (28%) open, 4 (11%) semi-open, and 22 (61%) close-ended questions (Appendix B, p. 207). The pages are organised to deal with similar research topics.

- *Page 1*: contains the abstract, and two questions about the respondent's biosphere reserve(s) and country.
- *Page 2*: includes eight questions that asks respondents to elaborate on the biosphere reserve's environment and its characteristics.
- *Page 3*: it contains nine questions focusing on the biosphere reserves' projects and their sustainability.
- *Page 4*: it has ten questions mainly relating to the potential impacts of recent international events on the biosphere reserve network, and the relationship-strength between biosphere reserves and the different network levels.
- *Page 5*: has four questions relating to the biosphere reserve's relationship with the local population, businesses, organisations and companies, the impact climate change will have on local livelihood, and the local level of climate change awareness.
- *Page 6*: contains three questions; the first relates to what biosphere reserves could do to improve their capacity to address climate change, and the other two once again relates to which scenarios respondents believes to be most likely to happen, and in which scenario their organisations' efforts fit.
- *Page 7*: thank respondents for completing the questionnaire.

The second questionnaire aims to obtain more details about certain elements mentioned in the first questionnaire. It therefore does not repeat the scenario format.

4.6.3 Reliability and validity

Reliability and viability of research instruments provide assurance that the study has accurate and replicable results, that follows the scientific process (Bryman, 2016).

4.6.3.1 Reliability

Reliability is the consistency of a measure in a research instrument. This means that if the same questions in a questionnaire are repeated using the same methods, under the same circumstance, to the same individuals the results will be consistent (Fink & Litwin, 1995). This is vital in order to draw accurate conclusions or even make predictions (Drost, 2011). This study employs two methods to test reliability, namely stability and internal reliability.

Stability is tested with the test and re-test method. This is when a test is administered and then readministered to the same sample at a different point in time (Bryman, 2016). In this research study there are three repeated questions, namely: "Do you think the UNESCO Biosphere Reserve status and its global network have a substantial impact on climate change adaption and mitigation?"; "Which climate change scenario do you think is most likely to happen?"; and "In which climate change scenario do your organisation's efforts fit?". To determine the stability from the first round to the second round of questionnaires, the correlation between the answers is determined by calculating the correlation coefficient (Fink & Litwin, 1995). A correlation coefficient is a statistical measure that quantifies the degree to which two variables are linearly related. In other words, as one variable changes, the other variable changes in a proportional and constant manner (Verma, 2019). The value of this coefficient ranges from -1 to 1; values between -1 and 0 describe an inverse linear relationship, while values between 0 and 1 describe a positive linear relationship. In the formula (Karros, 1997):

$$r = \frac{\Sigma(x_i - \underline{x})\left(y_i - \underline{y}\right)}{\sqrt{\Sigma(x_i - \underline{x})^2 \Sigma\left(y_i - \underline{y}\right)^2}}$$

- Σ represents the summation symbol (adding up all the values),
- x_i and y_i are the individual data points of the two variables,
- $\underline{x}$ and $\underline{y}$ are the means (average) of the respective variables.

When this formula is applied to the questions from the first questionnaire that were repeated in the second questionnaire, the correlations are as follows:

- Do you think the UNESCO Biosphere Reserve status and its global network have a substantial impact on climate change adaption and mitigation? = 0.99 correlation
- Which climate change scenario do you think is most likely to happen? = 0.98 correlation
- In which climate change scenario does your organisation's efforts fit? = 0.00 correlation

While the first two questions almost perfectly correlate, the last question has no correlation. This, however, is not because the question is unstable or unreliable, but rather because most respondents changed their minds and their answers over a six-month period, and that is what we aimed to determine. An example of other research on African biosphere reserves that used the correlation coefficient to establish reliability is: "Slenderness coefficient models for tree species in Omo Biosphere Reserve, South-western Nigeria" by Oladoye et al. (2020).

The second form of reliability used in this study is internal reliability. Internal reliability assesses whether the questions within the instrument or questionnaire measure the same underlying construct or idea. In other words, it evaluates how well the questions in a questionnaire interlink, indicating the extent to which they are all measuring the same concept (McDonald, 2002). Internal reliability can only be applied to a multi-item question (like a Likert scale) where the scores are aggregated to form an overall score (Bryman, 2016). In such a scenario it is possible that not all the questions in a questionnaire relate to the same overall concept and lack coherence between the questions. To determine whether a questionnaire has coherence, Cronbach's Alpha is often employed (Fink & Litwin, 1995).

Cronbach's Alpha is the most common tool to measure internal reliability as it measures the average correlation between all possible combinations of items (questions) on a scale (questionnaire). A high Cronbach's Alpha value (typically above 0.7 or 0.8) suggests an acceptable internal reliability, indicating that the items in the scale sufficiently correlates with each other and are likely measuring the same underlying concept (Tavakol & Dennick, 2011). Cronbach's Alpha is expressed in the following formula (Bujang et al., 2018):

$$\alpha = \frac{N}{N-1}\left(1 - \frac{\sum variance\ of\ individual\ items}{variance\ of\ the\ total\ score}\right)$$

- N is the number of items in the test,
- Σ represents summation,
- "Variance of individual items" is the variance of each item's scores across all participants,
- "Variance of the total score" is the variance of the total scores of all participants on the entire set of items.

This formula was applied to the questionnaires in the IBM Statistical Package for the Social Sciences (SPSS) and questions containing a Likert scale (Appendix C, p. 223). The Cronbach Alpha score achieved in both first and second questionnaires was: 0.73. This means that both questionnaires fall within the acceptable range of internal reliability, and that the range itself is consistent between the two questionnaires. An example of another research paper on an African biosphere reserve have used Cronbach's Alpha is: "Determinants of environmental conservation in Lake Tana Biosphere Reserve, Ethiopia" by Bires and Raj (2019).

4.6.3.2 Validity

Validity of a research instrument refers to whether an indicator (questions) meant to determine a certain concept really measures that concept. There are several ways to establish research instrument validity, namely content (is the test fully representative of what it aims to measure), face (does the content appear to be suitable to what it aims to test), and criterion (does the result accurately measure the concrete outcome it was designed to measure) (Schutt, 2009). This research, however, uses a fourth form of validity, namely construct validity, which aims to establish how well a questionnaire measures a construct that it was designed to measure (Strauss & Smith, 2009). A construct is a theoretical concept or theme based on empirical observations, and is not directly measurable (O'Leary-Kelly & Vokurka, 1998). In this instance the theoretical construct is the AfriMAB Biosphere Reserve network's social-ecological system. An example of other research in African Biosphere Reserves that has used construct validity is: "Charisma and conservation: charismatic megafauna's influence on safari and zoo tourists' pro-conservation behaviours" by Skibins et al. (2013).

To ensure that the research instrument is valid, a study needs to base the instrument on findings from its literature review and determine which possible questions could be asked to measure it (Dellinger, 2005). All the questions asked in both questionnaires relate to either elements in the literature review or the theoretical framework e.g., social-ecological systems, biosphere reserves, and/or climate change. Once the research instruments were prepared, experts were asked to evaluate the questionnaire and make sure the measures were appropriate, the questions clear, and unambiguous. This method also falls under face validity

(Messick, 1995). As the first questionnaire anchors the research inquiry and set the tone for the entire study, several experts in research, nature conservation, and biosphere reserves reviewed the questionnaire. Once these experts' comments were included and the questionnaire, an improved pilot study was conducted (Messick, 1995; Feeley et al., 2021). The first questionnaire was physically handed to nine of the participants of the "Towards climate neutrality in European UNESCO Biosphere Reserves workshop" on the Isle of Vilm on 1 April 2022. This questionnaire was subsequently distributed online to the 26 member AfriMAB youth network, of whom only three responded. These pilot-study participants highlighted ambiguous or unclear questions.

The second questionnaire builds upon or reiterates the questions of the initial one. This research tool was circulated in early 2023, following its submission to various experts at the Geography and Regional Science Department at Julius Maximilian University Würzburg. Their feedback was sought to enhance the questionnaire, and they were invited to provide their insights by both offering comments for improvement and completing the questionnaire themselves.

4.7 Sampling strategy

In this study we use questionnaires to collect data from a sample of the research population. The sample is the number of individuals within the participating group (Daniel, 2011). There are two primary forms of sampling, namely probability and non-probability sampling. Probability sampling involves a random selection of participants, and non-probability sampling involves purposefully selecting a sample (Bryman, 2016).

This study employs non-probability sampling, as candidates for answering the questionnaires need to have specific attributes, namely that they must be members of, or affiliated with, the UNESCO AfriMAB network, and must be adequately informed about these biosphere reserves and related climate change actions and be reachable via email. This network only has a finite number of members who qualify, necessitating a non-probability sampling avenue to better target them (Baxter et al., 2015). An example of a publication on African biosphere reserves that have used the non-probability sampling technique is: "Linking local people's perception of wildlife and conservation to livelihood and poaching alleviation: A case study of the Dja Biosphere Reserve, Cameroon" by Epanda et al. (2019).

Three forms of non-probability sampling are used in this study, namely voluntary response sampling, expert sampling, and snowball sampling. Using voluntary response sampling, the survey was distributed to the population of individuals who meet the above-mentioned criteria, inviting them to volunteer to participate (Padmavathy et al., 2023). This form of sampling is inexpensive and easily feasible, but its disadvantages include that there is a possibility of bias, and that the researcher has little control over the composition of the sample (Murairwa, 2015; Stratton, 2023). Another study on African biosphere reserves where voluntary response sampling is used, is: "Understanding communities' perceptions, demographics and uses of wetlands in Vhembe Biosphere Reserve, South Africa", by Dalu et al. (2022).

Expert sampling was used when the survey was distributed through the AfriMAB email list (dated 24 September 2020; obtained 7 March 2022), which mostly consists of people who meet the required criteria. This was the most up-to-date list at the time and contained 100 members. Once these individuals were contacted and responded their ability to answer the questions were unveiled. Because the questionnaire requires respondents to be experts on their specific biosphere reserves and projects relating to climate change in the AfriMAB network, this study has an element of expert sampling. The advantage of sampling experts is that the data they provide will be more specific to the research goal, but the disadvantage is that one can never be 100% sure about the respondent's level of expertise, especially with an online survey (Singh, 2009). An African study that employs expert sampling is: "Facilitating biodiversity conservation through partnerships to achieve transformative outcomes" by White et al. (2023).

Ten of the 100 members of the AfriMAB network that were contacted considered themselves inadequately equipped to participate in the study, but they suggested better suited candidates to take their place. This has an element of the snowball sampling technique, where research subjects recruit other research subjects to complete a research instrument (Naderifar et al., 2017). The advantage of the snowball sampling technique is that the survey reaches the most suitable respondents who can provide the most valuable information. The disadvantage is that it opens the survey up to increased sampling errors (Sadler et al., 2010). Another study of African biosphere reserves that used snowball sampling is "Assessing local people's perceptions and preference for ecosystem services to support management plan in Omo Biosphere Reserve, Nigeria" by Adeyemi et al. (2022).

4.7.1 Population and sample response

The first questionnaire was partially completed by 70 respondents, but only 56 completed it to the last page (Figure 4.3). The response rate was 70%, which is classified as acceptable by Fincham (2008). Africa boasts over 2,000 languages, highlighting its linguistic diversity. Given this variety, English and French were chosen for the questionnaire. These prominent international languages are spoken in a significant number of African countries: English in roughly 44% and French in around 39% (Diallo, 2018). This selection maximizes the likelihood that respondents can understand the questionnaire, compared to languages with a more limited reach, like Portuguese (spoken in about 21% of African countries) or

Figure 4.2 The response rate decline of the first questionnaire over a ten-week period.

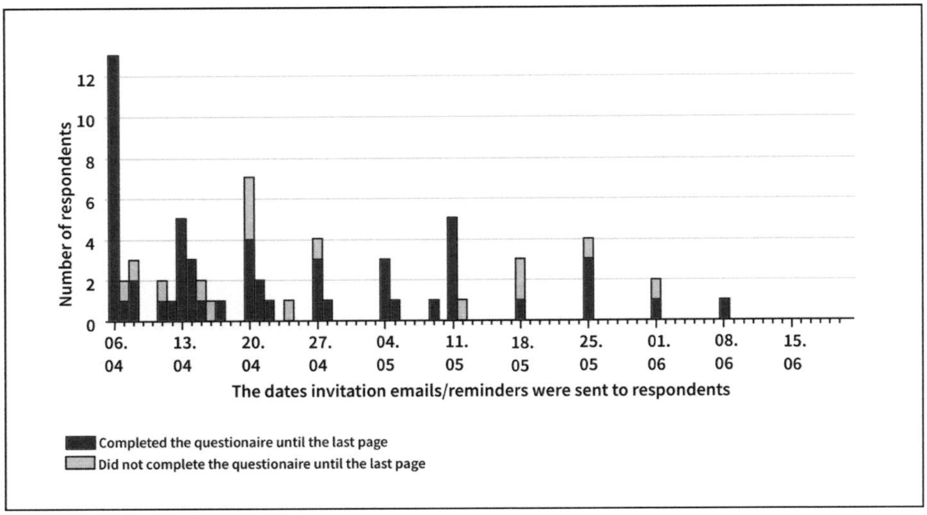

Source: Own compilation

Spanish (used in only about 2% of countries). Respondents therefore completed the questionnaire in English (44%) or French (56%). This survey was opened on 6 April 2022, and it ran for ten weeks, until 15 June 2022. Potential responders were reminded weekly to answer the questionnaire, and to complete it to the last page.

In the first questionnaire respondents were asked if they would participate in the second round of questionnaires. Of the 70 respondents to the first questionnaire, 63 agreed to participate; this made them the new survey population. From this group, only 38 participated in the second round, of whom 32 completed the questionnaire to the last page. The response rate for the second questionnaire was 60%, and even though it is less than the first round, it is still within the acceptable range (Fincham, 2008).

This survey was distributed on 19 January 2023 and was open for ten weeks until 30 March 2023, with weekly reminders asking respondents to participate or complete the questionnaire (Figure 4.4). The questionnaire could be completed in either English (53%) or French (47%).

Figure 4.3 The response rate decline of the second questionnaire over a ten-week period.

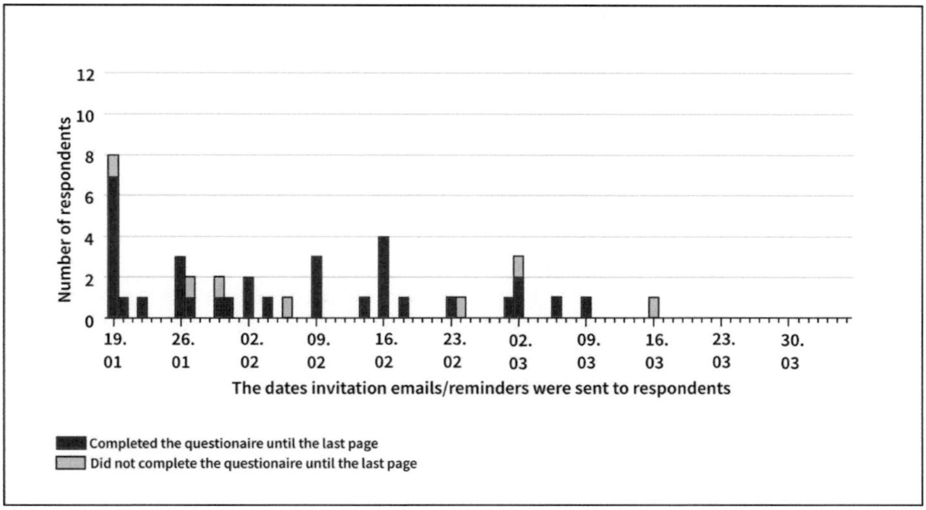

Source: Own compilation

4.7.2 Sampling bias

The sampling bias in a survey is the distortion of representativeness of the sample that occurs when members of a population has little to no chance of being included in the study (Bryman, 2016). While both surveys tried to reduce sampling bias, by giving the entire population, or the entire volunteered population, the opportunity to complete the survey, sampling bias is still present (Sedgwick, 2013).

There are two obvious forms of bias that emerge from this survey. Firstly, in both surveys self-selection/voluntary response bias is present because individuals who volunteer to participate in surveys have certain personality traits. Individuals without these personality traits do not participate and their opinions are therefore not surveyed, causing a sample bias (Kelker et al., 2020). Also, both surveys had non-respondents, causing non-response bias, which occurs when suitable individuals partially decline to participate, or drop out entirely (Reschiwati & Soeprijadi, 2018). The second survey was especially affected by this. A researcher can do nothing about either of these biases, as they are driven by individuals' freedom of choice, which should be respected (Bryman, 2016).

4.7.3 Ethical considerations

It is important to incorporate ethical considerations into research sampling as it protects both research subjects and the studies' scientific integrity (Cacciattolo, 2015). There are three ethical issues that were addressed during the study. The first is voluntary participation. The participants were free to either complete or not complete both surveys, and they had the freedom to withdraw at any point in the study (Ryan et al., 2005). In fact, respondents to the first survey were asked to volunteer to participate in the second round, and if they declined the request, they were not contacted again.

The second ethical consideration is informed consent. Respondents were informed about the study aims, objectives and any other relevant information, so that they would know what the information they provided would be used for (Mandal & Parija, 2014). During both rounds of the survey respondents were informed about the details of the study, both on the invitation email that contained the link to the survey, and on the first page of each survey. Research participants were therefore thoroughly informed.

The third ethical consideration was anonymity, which means that the researcher does not know the identity or any personal identifiable data of the participants (Wiles et al., 2008). Email addresses were added to the online soscisurvey.de program which then anonymises participants. The programme provides results under an identification number, but states whether respondents have opened or completed the questionnaire or not. Specific data relating to participants is kept anonymous, ensuring that answers provided by an individual could not be traced back to them.

4.7.4 Limitations

This research survey sample had to address or accept several limitations, not the least being the fact that the project was conceived during the Covid-19 pandemic and commenced in 2020/2021. Personal contact with research participants was restricted by international travelling laws, so to protect both the participants and the researcher, it was decided to rather use an online survey. Access to the AfriMAB network was also limited and the only way to contact prospective participants was through an email list that was not readily accessible; it had to be obtained through a benevolent member of the network. Furthermore, the only way to determine the level of expertise of the participating members of the AfriMAB network is to rely on the information they provide about themselves or of the respondent they nominated in their place. Due to issues with internet connectivity and language hurdles, information had to be taken in good faith.

4.8 Data analysis methods

Data analysis techniques are the methods used to process raw data obtained from survey answers, to be interpreted and analysed (Bryman, 2016). The questionnaires were distributed and collected using the online platform soscisurvey.de and the data was then imported into IBM SPSS for further processing. The French data was translated into English, using the online translation tool DeepL.com. Subsequently all questions relating to the same biosphere reserve or country were aggregated into a single entry per question, per biosphere reserve. The quantitative data is organised into tables; this includes a spatial variable which is either a country or biosphere reserve. The data is then analysed to describe the most and least frequently mentioned data points, and where applicable the highest, lowest, and average scores are determined. Similar research that includes quantitative data analysis is: "Opportunities and challenges for wildlife conservation: The case of Alatish National Park, Northwest Ethiopia" by Berhanu and Teshome (2018).

The qualitative data is organised into thematic categories gauged from the data, which is then also related to a spatial element, either a country or biosphere reserve. Thematic analysis aims to identify and understand themes within the data (Braun & Clarke, 2021). The advantage of using thematic analysis is its flexibility, so it can be used for a variety of data types. A major disadvantage of thematic analyses is that it is based on the subjective interpretation of the researcher (Joffe & Yardley, 2003). Yet, its advantages outweighed the disadvantages, and it is therefore deemed an appropriate research tool for the study. A similar study on biosphere reserves that uses a thematic analysis is: "Learning to live with social-ecological complexity: An interpretive analysis of learning in 11 UNESCO Biosphere Reserves" by Schultz et al. (2018).

Most of the quantitative and qualitative data is lastly subjected to spatial analysis. Spatial analysis is defined as a collection of methods, statistics and techniques that integrates concepts such as location and geographic context patterns with explanations of patterns and behaviours (Grekousis, 2020). This study utilised thematic maps, which are maps that convey natural or cultural features of an area, and also geographic variables, e.g., size of biosphere reserves, age of biosphere reserve, and number of implementation zones of climate change projects. Most of the data is expressed in maps which makes it possible to examine and interpret patterns and relationships found within the data, within their geographical context (Slocum et al., 2022). The advantages of using thematic maps include that it makes it possible to analyse the variables within a spatial context, and that it makes it easier to visualise the data. The disadvantage of this method is that the data is often distilled into an easily visualised metric, meaning the intricacies and detail of the collected dataset could be lost (Ślusarski & Jurkiewicz, 2019). Other researchers studying biosphere reserves in Africa have also used spatial analysis and thematic maps to analyse their research, for example "Analysis and prediction of land cover change of a biosphere reserve using geospatial tools: A case study of Omo Biosphere, Nigeria" by Adeyemi (2018).

4.9 Concluding remarks

This chapter's objective was to establish the steps to be taken to ensure that the research is reliable, valid, and replicable, and that it will contribute to the academic body of knowledge on climate change and biosphere reserves. The following section elaborates on characteristics of the AfriMAB network study area.

5 The AfriMAB study area

5.1 Preliminary remarks

This chapter aims to provide some geographical and environmental context for this study. The study focuses on the African continent, which encompasses an area of about 30,37 million km^2. In terms of the UNESCO MAB programme, Africa is divided into two distinct regions, namely the Arab Network for MAB (ArabMAB) and AfriMAB, which is the UNESCO MAB regional network for sub-Saharan Africa. The reason for this division is because the northern African countries share similar environmental, economic, and cultural characteristics with the Arabian nations, which differs from countries south of the Sahara (Bentahar, 2011).

The following sections elaborate on the contribution and expected impact of climate change on sub-Saharan African countries; and specifically, how the UNESCO AfriMAB guidelines suggest biosphere reserves address it.

5.2 Contribution and impact of climate change on sub-Saharan Africa

The entire African continent contributes about 3.9% of the global greenhouse gas emissions. However, it is expected that this continent, especially the sub-Saharan African region, will suffer the most because of the impacts of climate change (IPCC, 2022). This is because this region contains some of the world's poorest countries that still mainly depend on stable weather systems and environments to survive. Climate change has already started to interfere with these systems, which is troubling as around 80% of Africans south of the Sahara depend on rainfed agriculture. The negative effects on food security are already visible in many of these countries (Ayugi et al., 2022). The changing weather systems further threaten high population-density coastlines with erosion and floods. Around 56% of the coastlines in Benin, Côte d'Ivoire, Senegal, and Togo are already experiencing an average of two meters of erosion per year (UNFCCC, 2020).

Extreme weather events are another challenge which is expected to intensify. In 2022, at least 4,000 Africans died because of extreme weather events. For example, around 2,500 people died in Uganda because of drought and famine; 600 people died in Nigeria's worst flood in decades; and six unusually severe storms hit Southern African countries (including Madagascar and Mozambique), killing at least 890 people (Dune, 2022). Across Africa incidences of individuals suffering from heat stress are increasing, as people are unable to maintain a comfortable internal temperature, which in severe cases may cause death (UNFCCC, 2020). Of grave concern is also the expectation that people will increasingly suffer from climate-sensitive diseases (including malaria, meningitis, and cholera) that may spread to areas where they are not currently common. These climate-related impacts are expected to escalate over the next few years (UNFCCC, 2020).

While many other societies in the world face similar threats, sub-Saharan African countries are particularly vulnerable because of their low adaptive capacity, which is due to weak economies, political instability, and frail governing institutions (IMF, 2020; Thiel et al. 2022). The International Monetary Fund (IMF) reported in 2023 that the ten poorest countries in the world were all located in Africa, namely: South Sudan, Burundi, Central African Republic, Somalia, Democratic Republic of the Congo, Mozambique, Niger, Malawi, Chad, and Liberia (IMF, 2023). At the UN Climate Change Conference COP27 held in Sharm El-Sheikh in 2022, it was shown that seven of the ten countries that are most vulnerable to climate change are in Africa, namely South Sudan, Central African Republic, Somalia, the Democratic Republic of the Congo, Chad, Nigeria, and Ethiopia. Thus, five African countries that are amongst the ten poorest countries in the world are also on the list of countries most vulnerable to climate change. These countries were identified because of their predisposition to political instability, droughts, floods, diseases, and famine (Allan et al., 2022).

Another index, the ND Gain Index published by the University of Notre Dame (2021), ranks countries according to their vulnerability to climate change and other global challenges and their ability to improve their resilience. This index uses the ability to adapt over six sectors crucial for sustaining life as vulnerability indicators namely: food, water, health, ecosystem services, human habitat, and infrastructure. The lowest-ranking sub-Saharan African country (out of a global 185) with the least vulnerability and the highest resilience is Mauritius at rank 48; the median rank is 154 with countries like Nigeria and Mozambique; and the highest-ranking country with the most vulnerability in in the world is Chad at rank 185. According to this index, most sub-Saharan African countries are vulnerable to climate change and do not possess adequate resilience (Universty of Notre Dame, 2021) (Figure 5.1). Ironically, there seems to be an inverse relationship between the per capita CO_2 emissions and the countries' vulnerability rank. This is because CO_2 emissions generally indicate economic growth and some degree of political stability (Althor et al., 2016).

A way to reduce climate change vulnerability could therefore be to encourage green economic growth with lower carbon emissions that includes socially and environmentally sustainable development (Filho et al., 2016). UNESCO AfriMAB Biosphere Reserves would be ideal mechanisms to implement such activities and act as testing sites for the rest of the region (Reed & Price, 2019).

Figure 5.1 The 2021 CO_2 emissions per capita and ND Gain Index per country.

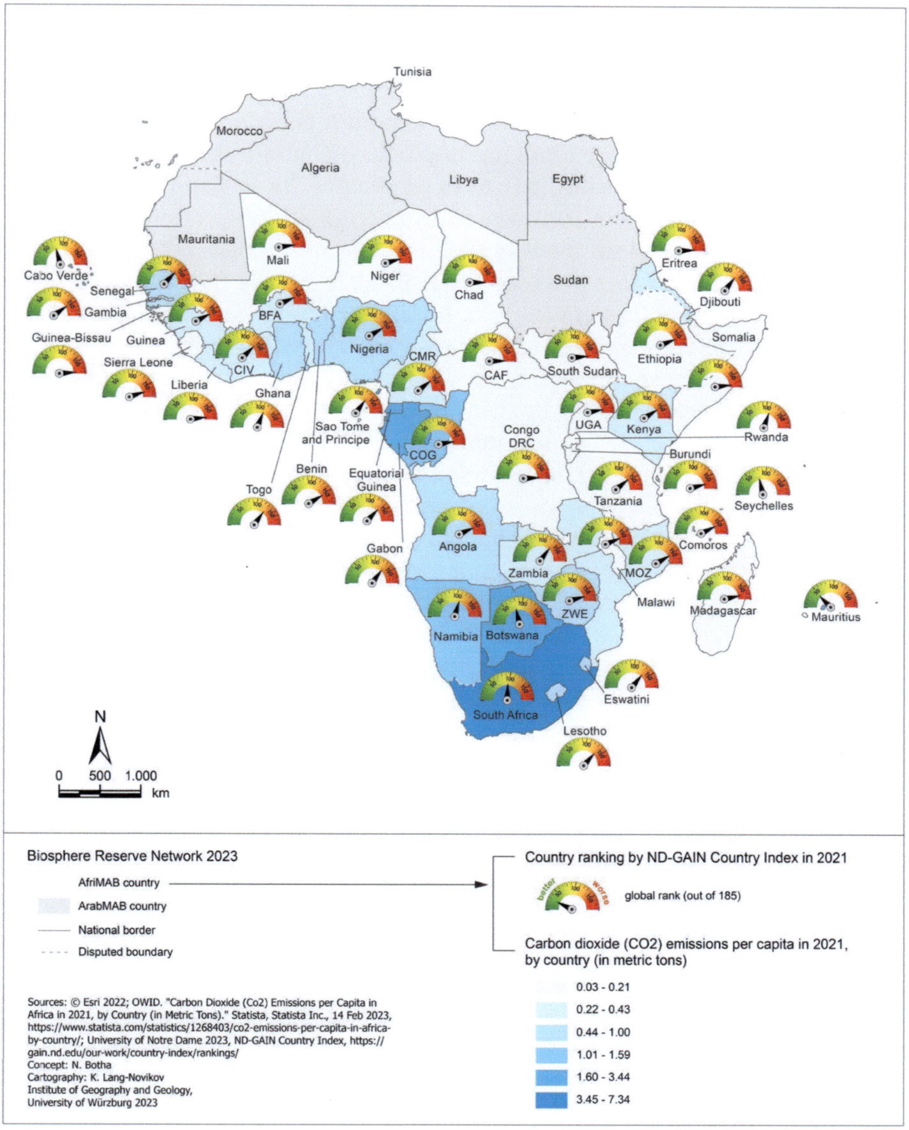

Source: Own compilation adapted from (Statista, 2021; University of Notre Dame, 2021; Esri, 2022)

5.3 AfriMAB and its guidelines to address climate change

The UNESCO AfriMAB Biosphere Reserve network was created in 1996 and is located in sub-Saharan Africa (UNESCO, 2018). It currently contains 93 biosphere reserves in 33 countries, covering a total of 1,027,585.58 km^2, or 4.2% of the sub-Saharan African region (24,328,298.87 km^2) (Figure 5.2). The smallest biosphere reserve is Samba Dia in Senegal at 8 km^2. The largest is Aïr et Ténéré in Niger, at around 239,700 km^2. There are more mountainous biosphere reserves in southern to eastern Africa, while western Africa is dominated by lowland biosphere reserves. The oldest biosphere reserves on the continent are scattered along the equator, likely because of early conservation research interest in equatorial ecosystems such as rainforests (White & Edwards, 2000). Many the older biosphere reserves tend to be smaller than the rest, but this might be because they were established before the implementation of the post-1995 biosphere reserve model.

AfriMAB aims to support and implement the guidelines of the Seville Strategy but has its own charter (AfriMAB, 2010) in which climate change is mentioned once under the heading of "Implementation strategy for the AfriMAB network". AfriMAB should enhance regional and international cooperation on climate change, desertification, and migratory species protection to support local communities in biosphere reserves. They should also focus on establishing South-South, and triangular North-South-South collaborations for sustainable development, prioritizing livelihood sustainability (AfriMAB, 2010).

The AfriMAB management manual addresses climate change in more detail as it is a key challenge (Amer et al., 2015) because it threatens sustainable development and poverty reduction efforts. Pre-existing challenges such as malnutrition, biodiversity and ecosystem services loss, desertification, urbanisation, water scarcity or pollution, is expected to be exacerbated by further climate change (Filho et al., 2021). While the impact of climate change is difficult to predict in Africa because of the divergent geographic and environmental contexts, the management manual suggests the following scenarios (Amer et al., 2015):

- *Decreased agricultural yields*: Rain-fed agriculture faces reduced yields, potentially up to 50% in some countries.
- *Infectious diseases*: May spread to highlands due to changing climate patterns.
- *Habitat loss*: Including expansion of arid areas, and effects on wetlands and aquatic ecosystems occur.

Figure 5.2 AfriMAB with biosphere reserve size, type, and age.

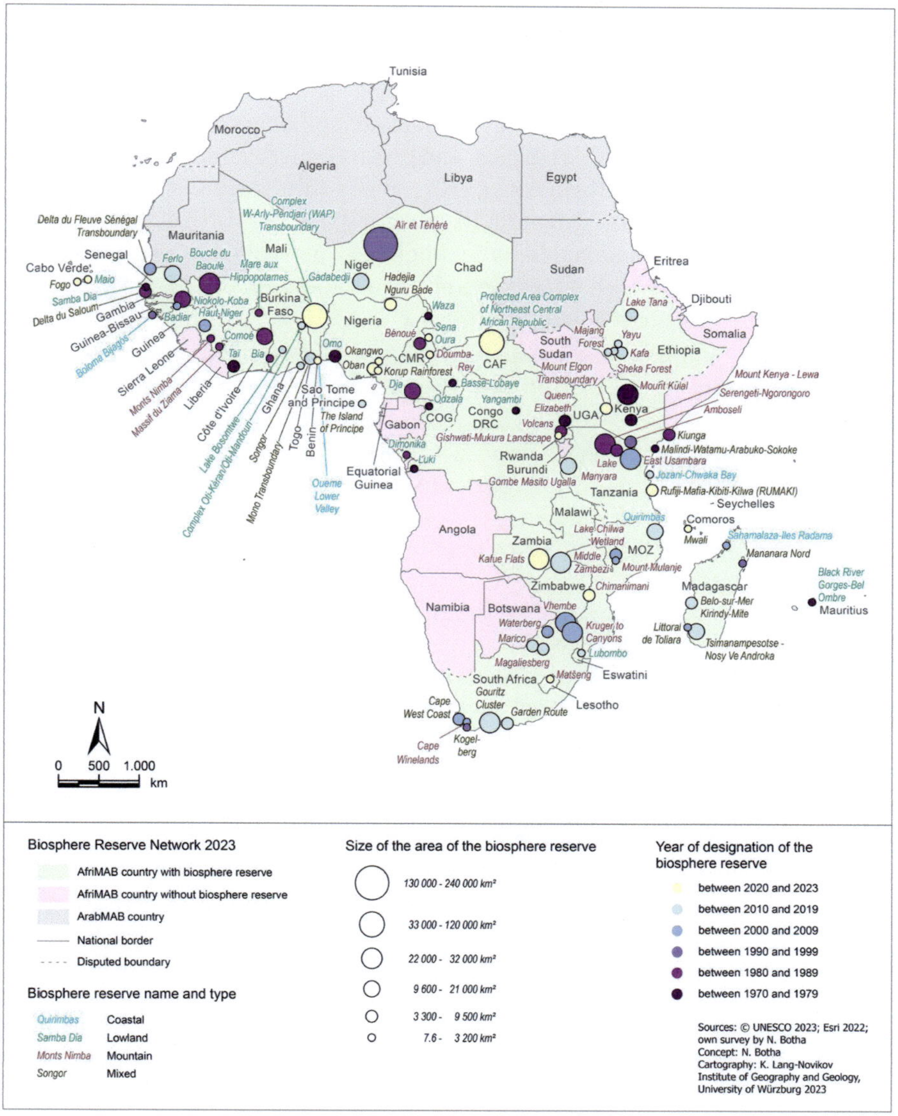

Source: Own compilation adapted from (Esri, 2022; UNESCO, 2023)

- *Biodiversity impact*: species face extinction, isolation, and reduced populations, especially in mountainous regions.
- *Invasive species*: may proliferate in changing ecosystems.
- *Human migration*: due to climate change may create a refugee problem (Amer et al., 2015).

AfriMAB has a limited potential for climate change adaptation as it currently focuses on land use, renewable energies, and green economy initiatives (Ad-daney et al., 2023). However, this manual suggests a focus shift from sustainable development mitigation strategies, to testing adaptation strategies (UNESCO, 2017). This includes protecting the sustainable livelihoods of vulnerable groups. Despite the suggested shift in focus, both adaptation and mitigation actions are suggested as potential activities (Amer et al., 2015; Reed & Price, 2019):

- Implementing measures to reduce emissions from deforestation and forest degradation (REDD+) or engaging in reforestation efforts.
- Preserving or restoring vital ecosystems such as mangroves, wetlands, and peat bogs, which store high levels of carbon.
- Adopting adaptive management strategies for rivers and floodplains to cope with anticipated increases in extreme weather events.
- Regulating slash-and-burn practices to minimize environmental impact.
- Promoting agricultural techniques that prevent erosion and help retain carbon in the soil.
- Abandoning shifting cultivation practices.
- Implementing measures to control sea erosion and shoreline recession.
- Monitoring saltwater intrusion into coastal aquifers.
- Introducing improved stoves in households using simple biomass.
- Utilizing sustainable sources of electricity.
- Enhancing water and energy efficiency in households.
- Generating income through a local tourism carbon credit mechanism, allowing tourists to offset their flight emissions by investing in local projects that mitigate climate change.

The impact of such activities depends almost entirely on the sustainability and efficiency of the institutional structures that are meant to implement these projects (UNESCO, 2017).

5.4 Concluding remarks

This chapter provided a geographical and environmental context for the study area. It was determined that AfriMAB is tasked with a monumental challenge in a region riddled with climate change vulnerabilities. Its task to find effective climate change solutions, with limited resources and constantly changing international goal posts, is a challenge not many other regions have to face. Hopefully this study will aid these biosphere reserves in their climate change adaptation and mitigation efforts.

6 AfriMAB social-ecological system framework results and discussion

6.1 Preliminary remarks

To determine how AfriMAB can position itself to improve its contribution to the CBD's 30X30 target, and specifically to address the impact of climate change, the results of two rounds of questionnaires that were sent to stakeholders were analysed. The collected data is grouped and discussed in the following categories of the social-ecological systems framework: Social, Economic, and Political Setting; Related Ecosystem; Resource System; Resource Units; Governance Systems; Actors; and Interactions and Outcomes. To discuss the results of the questionnaires, each question will be addressed separately.

6.2 Social, economic and political settings (S)

The social, economic, and political setting was established by investigating the following aspects: the level of respondents' UNESCO Biosphere Reserve network involvement; their contribution to the creation of any of the international climate change goals; and their attendance of recent international environmental goal setting conferences (Sharm El-Sheikh Conference, CBD, 2022). Additionally, they were asked about their perception of the impact of the UNESCO Biosphere Reserve status and its global network on climate change adaptation or mitigation.

6.2.1 At which level are you most involved with the UNESCO MAB Biosphere Reserve network? (n=63)

The results (Appendix D, Table 1, p. 227) indicate that the level of involvement of most of the respondents is local (77%), followed by national (75%), international (37%), and lastly other (8%). Locally, members could be part of a governance body of a specific biosphere reserve. On a national level, participants form part of government authorities or are members of the national UNESCO MAB networks. On an international level, members may for example participate in the ICC, or be associated with transboundary biosphere reserves. "Other" individuals do not belong to these categories but are involved with AfriMAB through their research or an external organisation that is closely affiliated with a specific UNESCO Biosphere Reserve.

On average, respondents are involved in their biosphere reserve network in at least two levels. The countries with the most representatives are South Africa (21%), followed by Côte d'Ivoire (14%) and Burkina Faso (10%). In South Africa (n=13), most representatives are local (92%) and then national (46%). In Côte d'Ivoire (n=9), most representatives are national (89%) followed by local (78%) representation. Burkina Faso representatives (n=6) are almost equally distributed between national (100%), international (83%), and local (83%). No country had more than one representative in the "other" category.

From these results, it is evident that local and national AfriMAB stakeholders are the two biggest groups represented in the questionnaire, and most respondents belong to both groups. Of the total number of respondents, almost 50% are from South Africa, Côte d'Ivoire, and Burkina Faso. According to Statista (2021) & University of Notre Dame (2021) South Africa is ranked 95 on the list of countries vulnerable to climate change; it has a lower-middle income; and emits 7.3 tons of CO_2 per capita per year. Côte d'Ivoire is ranked 140; is also a lower middle-income country; but only emits 0.4 tons of CO_2 per capita per year. This inverse relationship between economic strength and CO_2 production compared to South Africa raises questions about Côte d'Ivoire's economic production. According to the CIA Factbook, Côte d'Ivoire is one of West Africa's most influential, stable, and rapidly developing countries, mainly because of its cacao export and production, as well as foreign investment (CIA, 2023). Burkina Faso on the other hand, is ranked at 161 on the climate change vulnerability index; it is a low-income country; and only emits 0.3 tons of greenhouse gas emissions per capita per year (Statista, 2021; University of Notre Dame, 2021).

From these findings, it seems that the more vulnerable the country, and lower the economic activity, the fewer AfriMAB participants they have. Overall, these representatives of the AfriMAB network follow the guidelines and frameworks created by the ICC, which in turn are based on, for example, international environmental agreements such as the CBD's 30X30 target.

6.2.2 Were you involved with the creation of any of the previous international climate change goals? (n=25)

Only two respondents – one from Burkina Faso and one from Malawi – confirmed that they were involved in the creation of any of previous international climate change goals.

A possible reason for so little AfriMAB involvement in international climate change goals could be because of the sheer number of international intergovernmental organisations, programs, funds, and specialized agencies which each has its own agenda, conferences and other events that are scattered all over the world at varying time spans, as shown in the second chapter. Attending each of these meetings as a member state is expensive, especially for countries in sub-Saharan Africa that typically have limited resources. These intergovernmental organisations and their affiliated programs, funds, and agencies should ideally be analysed to ensure their core agendas are not overlapping, they should cooperate to reduce the number of conferences, and synthesise the key messages that they wish to deliver to member states. Reducing intergovernmental organisation redundancies might not only help poorer countries to attend and contribute to international events, but it will also strengthen the messages of these agencies and encourage the implementation thereof by member states. While Kotschy et al. (2015) suggest that such redundancies help improve an organisation's resilience, redundancy must be considered on a positive-negative spectrum, where at a certain point the redundancy's cost outweighs its benefits. In the context of this research, the UN's redundancies seem to result in a high cost of participation for member states that prevents lower-income countries from participating in international target-setting events. It is vital for the adaptive capacity of the World Network of Biosphere Reserves to broaden the participation of more vulnerable groups. Unfortunately, a lack of participation was also seen at more recent intergovernmental meetings.

6.2.3 Did you attend Sharm El-Sheikh conference on climate change in November 2022? (n=32)

None of the respondents attended this conference and most reported that they did not know it was held or what was discussed. This conference was organized by the UNFCCC also named COP27, a conference dedicated to addressing climate change challenges, in Sharm El-Sheikh, Egypt, from November 6 to 18, 2022. A reported 35,000 delegates from 198 countries attended the conference. Given that the focus of this conference was on climate change; that climate change disproportionately affects African nations; and the fact that the conference was hosted by an African country, one would have expected a higher attendance and greater awareness among respondents to this questionnaire. This conference aimed to establish a practical implementation plan to achieve the goals of achieving the Paris Agreement (UNFCCC, 2023). This implies that while these African countries might have been represented, it was not by local or national representatives associated with the AfriMAB network. This is unfortunate as they would have been ideal candidates to contribute to possible ways to implement climate change action on the ground.

The representatives from sub-Saharan African countries who attended COP27 apparently did not convey this conference's existence or information about it to those involved with the AfriMAB network. This raises the question: who speaks on behalf of AfriMAB? Are the individuals who attend these conferences suitably qualified to truly represent the specific interests of this network, and then to communicate the information that they obtained at the conference to the relevant stakeholders? Government communication is crucial for biosphere reserves to function effectively, and this is also what Schliep and Stoll-Kleemann (2010) found. It is possible that more recent conferences on climate change are deemed unsatisfactory (Isaac, 2021) because they are attended by too many people, causing the free-rider problem. To solve this problem, the UN and its member states should reconsider their criteria for delegate selection.

6.2.4 Did you attend the 15th Conference of the Parties (COP15) to the CBD in December 2022? (n=32)

A month after COP27, COP 15 was held from 7-9 December 2022 in Montreal, Canada, hosted by the CBD, specifically to finalise and agree on global nature targets for 2030 and 2050. The conference was attended by 15,723 delegates from 192 countries, including four of this study's questionnaire participants, who are from Cameroon, Côte d'Ivoire, Ethiopia, and Niger. The higher attendance of COP15 could be ascribed to any of a variety of reasons. It could support the theory that UN conferences suffer from the free-rider problem, as this conference had 43% fewer delegates, but a higher AfriMAB representation. Alternatively, it could mean that the CBD better communicates with member states than the UNFCCC. It could also be that the AfriMAB network is more concerned about biodiversity issues than climate change. Another possibility is that attendees were funded to attend the conference. The higher attendance rate may even be attributed to a combination of these factors. However, this issue is beyond the scope of this study, and to pinpoint the exact reasons for this phenomenon would therefore require further research.

6.2.5 Does the UNESCO Biosphere Reserve status and its global network substantially impact climate change adaptation and mitigation?

In the first questionnaire, respondents were asked if they thought that the UNESCO Biosphere Reserve status and its global network substantially impact climate change adaptation and mitigation (Appendix D, Table 2, p. 228; n=44). This question was repeated in the second questionnaire to determine if the respondents' opinions have changed over the six-month period between the questionnaires (Appendix D, Table 3, p. 228; n=22). Most stated that they think UNESCO Biosphere Reserve status substantially impacts climate change adaptation and mitigation (first questionnaire, 73% and second questionnaire, 68%) (Figure 6.1). Some (first questionnaire, 18% and second questionnaire, 14%) believed it potentially has a substantial impact, while others (first questionnaire, 7% and second questionnaire, 14%) think there has been no impact (Figure 6.1). In terms of percentages, the biggest change was the percentage of respondents who believe it does not have an impact (7%), but in terms of the number of people who stated it, it remained the same. Generally, there was a 4-7% deviation from the original questionnaire responses in each category.

Respondents were asked to provide reasons for their answers. Those who believed that the UNESCO status does contribute to climate change adaptation or mitigation refer to either single or several core functions of biosphere reserves as reasons for this, for example: "biodiversity conservation, socio-economic development, research, monitoring, training and public awareness". Some respondents elaborated on these functions: "biosphere reserves are sites of excellence for testing innovative approaches to sustainable development through scientific knowledge and governance methods. These functions are combined to reduce biodiversity loss, improve people's livelihoods, and foster the social, economic, and cultural conditions essential for a sustainable development model".

Other respondents referred to biosphere reserve activities: "they contribute to the protection of biodiversity, the mobilisation of knowledge on climate change, cooperation between different partners, the involvement of local communities in the management of the reserve, etc". One respondent commented on the role of the biosphere reserve's label: "it gives greater visibility to the reserve which is increasingly receiving special attention and support from all stakeholders facilitating their protection. This results in less and less CO_2 being released through the degradation of the environment". Because of this visibility a different respondent stated: "there is a lot of monitoring done to prevent people from destroying the core area and the buffer zone. Similarly, education and farming methods are being taught to the local population".

In one answer the biosphere reserves' network and its international associations were mentioned; another stated that the biosphere reserves' "international scale benefit from greater attention"; while yet another believes the main contributing factors to climate change mitigation and adaptation is the "the creation of a framework for meetings between all the stakeholders in the biosphere reserve area; travel and meetings to exchange experiences; organization of AfriMAB general assemblies; joint meetings of regional networks; participation in ICC sessions." This addresses climate change mitigation and adaptation "in terms of learning from other biosphere reserves in different network regions" and through participation in "projects/programmes such as MAB, IHP, IGCP, etc". This is because "mitigation and adaptation to climate change and other aspects of global environmental change is one of the MAB's Strategic Objectives for 2015-2025". A respondent from a relatively new biosphere reserve had some doubts, as they believe that "it does have a substantial impact if the overall MAB programme is considered. Individual biosphere reserves might not contribute as much, but this might increase as the biosphere reserves grow and gain support. I think older biosphere reserves will at this stage show better impacts on climate change adaptation and mitigation".

Figure 6.1 Comparing first and second questionnaire responses to whether the UNESCO Biosphere Reserve status and its global network substantially impact climate change adaptation and mitigation.

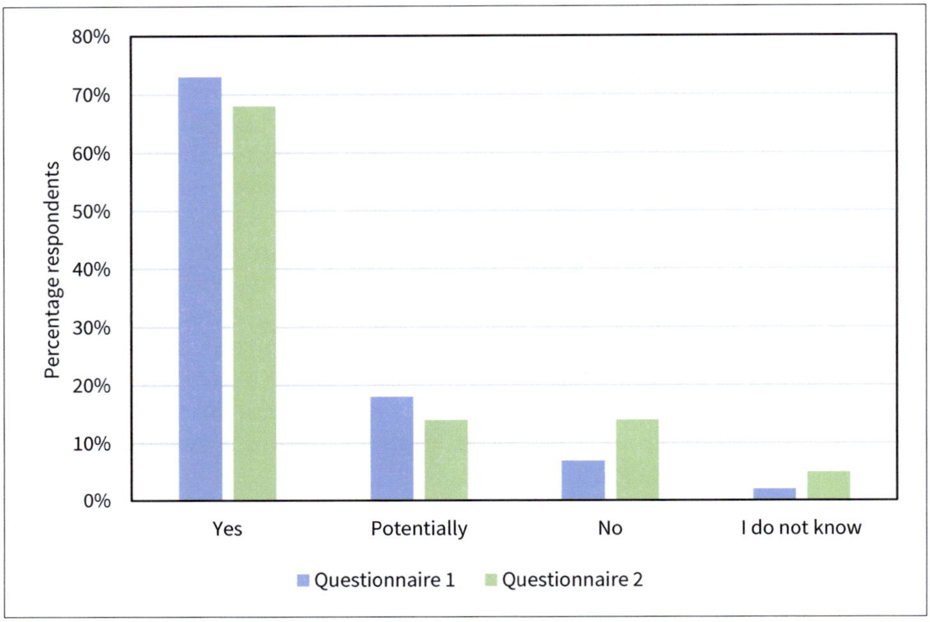

Source: Own compilation

Other explanations from respondents included: "biosphere reserves play fundamental roles in mitigating climate change, regulating and stabilizing global carbon, and storing large quantities of carbon over a long period of time by means of carbon sequestration into plant biomass which can enhance sustainable management". Another stated that "they help to maintain a microclimate favourable to the preservation of biodiversity and its development." The explanation given for this is that "the multiple (ecosystem) services of the biosphere reserve contribute significantly to climate change adaptation and mitigation".

In addition, one respondent stated that "the biosphere reserve status is a form of assurance that nature conservation can be sustainable. In terms of actions, it is more reassuring to invest in a biosphere reserve that is better protected legally and by actions that are more sustainable". Another said that "the conditionalities for preserving the status contribute greatly to the maintenance of the reserve's forest base". A different respondent believed that "it is a label that encourages the conservation of biodiversity and therefore of carbon sinks".

Community involvement was also mentioned: "the management of these reserves must largely involve the communities whose awareness of the problems of climate change and their commitment to face them or adapt to them through change, contributes greatly". In a similar vein, one respondent mentioned that "in my country and in the biosphere reserve it is the impoverished citizens who make the greatest impact on biodiversity loss and are the most vulnerable to the effects of climate change. The biosphere reserve offers an opportunity for the improvement of livelihoods and reduced dependence on biodiversity. It has already been shown to improve livelihoods and hopefully resilience to the impacts of climate change". Most other statements referred to need for community awareness about climate change.

An important explanation provided by respondents as to why they thought UNESCO Biosphere Reserves contribute to climate change mitigation and adaptation is funding: "it makes it easier for biosphere reserves to get funding on climate change related projects", and "through UNESCO's support to funding agencies for projects to combat the effects of climate change". However, this was also stated by respondents who believed the UNESCO Biosphere Reserve status and its global network potentially have a substantial impact on climate change adaptation and mitigation, as well as those who believe they do not.

A respondent who believed that the UNESCO Biosphere Reserve status and its global network potentially had an impact said: "in general the impact on climate change adaptation and mitigation through UNESCO Biosphere Reserves is good but the big issue is to change the mindset of local communities. This is difficult because most projects do well when it has finances but after the project ends (no finances) all activities stop, thus the status of UNESCO Biosphere Reserve has a substantial effect when there are finances to run projects". Another explained that "there is a need for long-term strategic support - both financial and technical - for biosphere reserve managers to take up a leadership and coordination role for research, planning, development of appropriate interventions, and coordination of action". According to one respondent "it has an impact, but not substantially ... my guess. This has to do with awareness. People need to know about biospheres and what it stands for. Locally this costs money, of which there is little ..."

One respondent described the potential of biosphere reserves in detail: "climate change requires constant adaptation to changing circumstances and to a changing biosphere. Interdisciplinary research on crosscutting themes across academic disciplines is needed as well as very specific research on a global scale and in a long-term comparison. The climate change related knowledge collected so far must be constantly reviewed and updated. People's awareness of climate change and its consequences within the biosphere reserve should rise significantly. This includes mutual learning between all stakeholders. Effective climate change mitigation and the development of adaptation strategies are the key challenges of environmental policy these days. (...) UNESCO Biosphere Reserves are ideal spaces to set up and improve such comprehensive learning processes in the context of climate change – a challenge for all of society. Biosphere reserves exist worldwide, they have appropriate management structures, and they have appropriate mission statements. It is about time to make better use of this instrument".

Other respondents stated that they believe the UNESCO Biosphere Reserve status and its global network is an "underutilised tool, there is too little data, large-scale projects are needed, and that more time is needed for the young biosphere reserve to mature".

Respondents who expressed some doubt about whether the UNESCO Biosphere Reserve status and its global network contribute to climate change actions, stated that the main reason was funding, or a lack thereof. "At this stage, there is very little impact related to climate change adaptation and mitigation. In the revised version of the Congo's 'Nationally determined contribution to the reduction of greenhouse gas emissions' document, there are projects related to climate change, but funding is needed". Another simply stated that there is a: "lack of funding on the ground". In a different vein, one respondent stated that: "the biosphere does not have a substantial impact - it depends on individual farmers and government organisations".

Overall, respondents mostly believed UNESCO status contributes to climate change adaptation and mitigation and most of those reasons are tied to the post-Seville biosphere reserve model and targets as described by UNESCO (2020). This supports the findings of Cuong et al. (2017b). Respondents identified two main areas where the status does not contribute, namely in long-term project support and sustainability; and local community support and awareness. These aspects were also identified in Brenner and Job (2022) and Botha (2023). Projects should therefore be planned with long-term implementation in mind, and the UNESCO MAB programme should enhance its identity and visibility.

6.3 Social, economic and political settings (S): Future prediction

The perception of the success or failure of an undertaking often becomes a reality as people tend to act according to those perceptions, making it a self-fulfilling prophecy. It is thus important to establish these viewpoints, especially in reaction to international agreements on which the UNESCO MAB goals are based. Respondents were therefore asked to predict the impact of the latest intergovernmental environmental conferences held in 2022.

6.3.1 Do you foresee any changes that the Sharm El-Sheikh Conference 2022 might bring about in the UNESCO Biosphere Reserves? (n=21)

The majority (9) of respondents were not aware of the conference or what was discussed. These respondents were from Burkina Faso, the Democratic Republic of the Congo (DRC), Côte d'Ivoire, and South Africa. As one respondent from South Africa stated: "it is difficult for me to say that maybe those meetings' outcomes will have an impact because we did not even know of those meetings. I think maybe the people who were responsible for inviting biospheres did not do a good job of transferring the outcomes of the meeting. I think it's something that is lacking because even in the best way, I think it will be good to be informed about such outcomes because those outcomes can also inform some of our research. What is it that we need to focus on so that we do research that is aligned with what is happening at an international level? Because if you're talking about climate change, you see that most of the challenges that we are facing are cross-cutting, whether you're in North Africa, or Asia. I think it's based on communication and there is generally poor communication".

Several respondents (7) predicted that this conference will have a positive impact. A respondent from South Africa predicted that: "more community climate change projects will emerge". Another South African said: "that there will be an urgency for a transformation in the economy to adopt green technologies, without impacting those reliant on fossil fuels". A respondent from Rwanda predicted "that there will be more funding for loss and damage around biosphere reserves". Four respondents were cautiously optimistic, with two, one from Cameroon the other from Niger, stating that the outcomes of this conference will only have a positive impact on the biosphere reserve if all state parties honour their agreements. One respondent from Mali stated, "that it could potentially have a positive effect if decision-makers invest more in the biosphere reserve", and another from Ethiopia simply stated that "it is an ambitious plan".

A few respondents (5) from South Africa, Burkino Faso, and Côte d'Ivoire did not think that the conference will make a difference. Most answers were simply "no", but a respondent from Eswatini elaborated: "I do not see any changes, at least not in the immediate future. The Southern Africa region is beset by an energy crisis, and this is impacting industries, livelihoods, and well-being. The phasing out of fossil fuel power plants will therefore not be immediate especially because most clean energy production processes are still at a developmental stage. Implementation of the agreements from this conference will raise the price of energy which will impact the price of goods and thus put more pressure on ecosystems".

It is surprising that the respondents from the three most representative countries in the AfriMAB network were not even aware of the conference, or what was discussed. The impact predictions of this conference are therefore limited. Most predictions either relate to greening the economy or improved funding.

6.3.2 What do you think of the decision at COP15 to try and protect 30% of the biodiversity of land and oceans by 2030? (n=28)

Most respondents (19) stated that they support this goal and that they think it is achievable. One South African said that it is "a good target, which will require considerable collaboration and funding to be achieved". Another from Burkina Faso commented that "if countries were to do this, it would be of real benefit to climate change mitigation". A respondent from Eswatini said that "it is a better target than the previous one however I would wish for an increase in this. If more than 75% of land and sea were treated as biosphere reserves, then it would be better. In other words, we need to treat every available piece of land and ocean as protected and only use resources in a sustainable manner". Another respondent, from Côte d'Ivoire, agreed stating that: "the target is achievable if the dimension of improving socio-economic development and logistical support is included, in short, the promotion of the biosphere reserve concept". In addition, one respondent from Rwanda stated that: "this needs to be implemented with full collaboration of communities so that governments are not just taking over lands for restoration". Another respondent, from Cameroon, noted that it is a realistic target, if "equity is respected in the sense of polluter pays in climate justice".

Six people were somewhat optimistic about the targets. A respondent from Burkina Faso thought that "this is a difficult goal to achieve. But not impossible". A respondent from the DRC thought "this is a good vision, but it requires substantial resources". Another from Côte d'Ivoire said: "this target can be met if the commitments made are respected". A Malian agreed with the target but felt "it would be more desirable to ensure good management of the old reserves already created". Others mentioned challenges that they might face in implementing the targets. A respondent from Burkina Faso mentioned "war and other upheavals" and a South African said that they: "have to deal with development and a growing economy".

Three respondents did believe the targets are achievable. A Tanzanian said: "it is impossible due to scarcity of resources", and another said: "on paper that percentage is achievable in South Africa. However, realising and enforcing that conservation target (i.e., really protecting the land and biodiversity) is not realistic". A Cameroonian commented "the means to do so exist but it does not seem to me that the political will among the stakeholders is there".

Overall, a consensus of 68% was reached as respondents believed this will be an achievable and positive target. The respondents who were not as convinced about the target, were all concerned about either the practical implementation of the target, or its sustainability.

6.3.3 What would be the impact of the 30X30 target on the UNESCO Biosphere Reserves you are familiar with? (n=24)

Overall, respondents (16) were positive about the impact of the targets. Five respondents stated that they predict the size and/or composition of the biosphere reserves will change. For example, a respondent from Malawi said they "are working towards a Malawi-Mozambique Transboundary Biosphere Reserve that will have many core zones". Another from Rwanda predicted the "expansion of parks, and hopefully improved buffer zones", while a stakeholder from Niger remarked that this target "will primarily concern biosphere reserves insofar as, as their name indicates, biosphere reserves at their core constitute sanctuaries or reservoirs for the conservation of biodiversity. Consequently, more than any other structure, biosphere reserves are tools for managing, conserving, and enhancing biodiversity".

Others mentioned the positive impact this will have on the environment and society. A respondent from Burkina Faso said that "this objective will increase biodiversity. It will also improve the resilience of populations". Another stated that this target will enhance the "preservation of endangered species", while a respondent from an unknown country predicted "livelihood improvement because if we take care of our nature more will be benefited from it". A respondent from the DRC stated that "The resilience of local people to climate change will be enhanced. The pressure on the reserve's resources will be reduced, the population will be aware of the benefits of the biosphere reserve's existence".

Eight respondents were less optimistic and doubted the targets can be achieved. A Malian stated: "if achieved and applied to the required technical standards, it can help mitigate the adverse effects of climate change to a greater extent". A Nigerian said that "it will be a remarkable achievement for age-long conferences", while a South African felt "it will largely depend on whether the promised funds actually trickle down to the biosphere reserves or stay at a higher level (with no practical impact)". Somebody supported this, saying not much will change: "on a local level we strive for lofty figures, but without money it is all wishful thinking". Another South African stated that: "policy and legislation will need to change to realise 30X30 targets in biosphere reserves. This will take years". Yet another South African noted: "I do not think we'll achieve this target. But it's better to have a target, even if you do not get there, even if you can achieve 20%, that's even better".

It is therefore predicted that the outcomes of this conference will cause an expansion of core and buffer zones positively effecting landscape conservation and biodiversity protection. This is in line with what the 30X30 target hopes to achieve (Dudley & Stolton, 2022).

6.4 Related ecosystems (ECO)

6.4.1 With which thematic network within the world network of biosphere reserves are you most associated? (n=20)

The UNESCO MAB network is divided into seven thematic sub-networks with 20 participating countries (Figure 6.2). On country level biosphere reserves participate, on average, in three different UNESCO MAB thematic networks.

The thematic networks that have the most participants are the wetland (60%), and savanna (60%), followed by the tropical forest (50%) (Appendix D, Table 4, p. 229). The contributing countries with the largest variety of thematic network participation are Côte d'Ivoire (7) and Mozambique (7). They are followed by South Africa (6), as it does not have a mangrove network representation. These countries are therefore the most diverse in terms of the biomes they contain.

The biomes in AfriMAB tend to cluster in specific areas: drylands occur mostly in western Africa; mangroves and tropical biomes are more common around the equator. Marine, coastal and island biomes are obviously limited to coastal countries. Mountain biomes are mostly situated on the eastern side of the continent, while savanna biomes seem to occur more frequently in the west. Wetlands are spread throughout the AfriMAB region, and often occur in the same countries as drylands.

The thematic network participation of each country provides us with insights into their location and environmental context, allowing us to consider the possible contributions the environment might make to adaption to, and mitigation of climate change. For example, the top three most represented thematic networks have the potential to help mitigate climate change. The wetland network is vital for the carbon sequestration and long-term storage of carbon dioxide (Mitsch et al., 2013). The tropical forest network has a critical role as carbon sinks, but because of deforestation, logging, and burning, their capabilities to perform carbon sequestration has been greatly diminished (Mitchard, 2018). It has been found that the savanna biome can capture as much CO_2 as tropical forests if they are managed well, but this biome's contribution is often overlooked (Dobson et al., 2022). AfriMAB should therefore focus on the restoration and protection of these key biomes to help mitigate climate change, which is once again the aim and focus of the CBD's 30X30 target.

Figure 6.2 AfriMAB countries' thematic network participation.

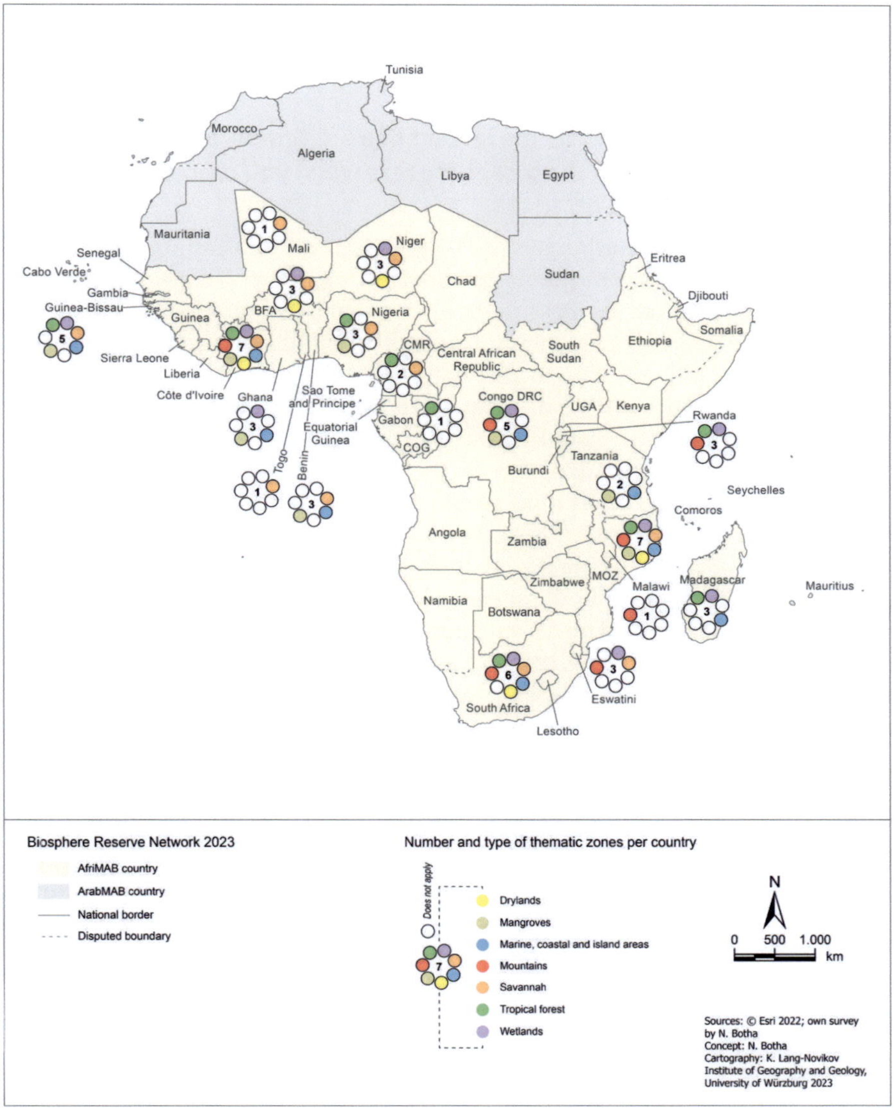

Source: Own compilation

6.5 Related ecosystem: Future prediction

The future prediction of the related ecosystem was determined by asking what the possible impact of a global 2°C temperature increase on the different zones of the biosphere reserve, and what temperature increase scenario respondents believe is most likely to happen.

6.5.1 What do you think will be the impact of a 2°C temperature increase on the different zones of the biosphere reserve(s) you represent or have intimate knowledge of? (n=14)

The answers to this question (Appendix D, Table 5, p. 230 and Table 6, p. 232) indicate that at least 50% of the respondents expected that there will be an increase in drought, 43% expected an increase in extreme events, 20% predicted a change in microclimates causing species migration, and 7% foresaw coastal erosion. The changes that are predicted due to fluctuations in the climatic conditions differ from zone to zone. In the core zone, most respondents mentioned biodiversity loss (71%) and increased water-related issues (e.g., drought and floods) (57%), and also habitat loss (50%) as their main concerns. Other threats that they mentioned were species migration (29%), anthropocentric pressure on the natural landscape (29%), increased instances of wildfires (14%), and natural resource loss (14%).

An increased variety of threats were mentioned in the buffer zone, with biodiversity loss (29%), water-related issues (29%) and habitat loss (29%) once again predicted most frequently. This was followed by predictions of natural resource loss (21%), increased anthropocentric pressure (14%), and increased human-nature/wildlife conflicts (14%). Other changes that were predicted include food shortages (7%), increased number of wildfires (7%), and poverty (7%).

In the transition zones, there was a larger variety of possible threats predicted, but the most frequently mentioned concern remains biodiversity loss (50%). This was followed by an increase in water issues (36%), and anthropocentric pressure on the natural landscape (29%). Respondents also predicted habitat loss (21%) and an increase in poverty (21%) because of communities' dependence on the natural environment. They were further concerned about increased natural resource loss (14%), an increase in wildfires (14%), an increase in human-wildlife/nature conflicts (14%), population migration (14%), cultural changes (14%), soil impoverishment (14%), and desertification (14%). The respondents seemed to be the least concerned about species migration (7%), housing issues (7%) and an increase in diseases (7%).

These predictions indicate that biodiversity and ecosystems services loss are the main climate change concerns of the AfriMAB community. These challenges cause knock-on issues in the buffer and transition zones, as droughts and extreme events cause microclimate changes leading to biodiversity loss and changes in ecosystem services (e.g., provisioning, regulating, cultural and supporting services). The reduction in these services (e.g., food, fuel, and water supply) will consequently encourage wildlife migration away from the core areas, as it can no longer support them. In addition, with humans encroaching on wildlife habitats to utilise the remaining ecosystem services, human-wildlife conflicts will emerge. This increased pressure on remaining resources will lead to food shortages and poverty in the buffer zones. This in turn will cause people living in the rural areas to either exhaust the natural resources that they have, or alternatively to move to the urban areas in the transition zone in the hope of finding a better livelihood. Consequently, this will cause cultural shifts, housing issues, and, because a large population is now concentrated in one area, the transmission and spread of infectious diseases.

To prevent most of these predicted negative impacts of climate change, the AfriMAB network should therefore focus their attention on efforts to protect biodiversity and enhance ecosystem services in the core and buffer zones. To accomplish this, the CBD's Kunming-Montreal Global Biodiversity Framework and the subsequent 30X30 target are an ideal approach to address the climate change issues within the region. The problem however is that the CBDs global biodiversity 10-year target setting conference is held five years before the World Network Conference for Biosphere Reserves. This conference is where the biosphere reserves' ten-year targets are updated to include "new" international targets such as 30X30. This means that under normal circumstances biosphere reserves only have five years to accomplish targets that were set for a ten-year period. This is a steep time frame, especially for less affluent countries, and the inclusion of CBD targets in the biosphere reserve framework is therefore more symbolic than practical. This lack of participation in striving to reach the set targets diminishes the usefulness of biosphere reserves as a tool for climate change action. The UNESCO MAB programme may therefore consider either moving the World Network Conference for Biosphere Reserves closer to the start of the decade or have more frequent conferences to encourage the acceleration of the incorporation of new intergovernmental targets into the biosphere reserve structure.

6.5.2 Which climate change scenario do you think is most likely to happen?

In this question, respondents were presented with three different scenarios, namely scenario one, where global temperatures increase with more than 2°C by 2050; scenario two, where global temperature increases remain below 2°C by

Figure 6.3 Comparison of predicted climate change scenarios.

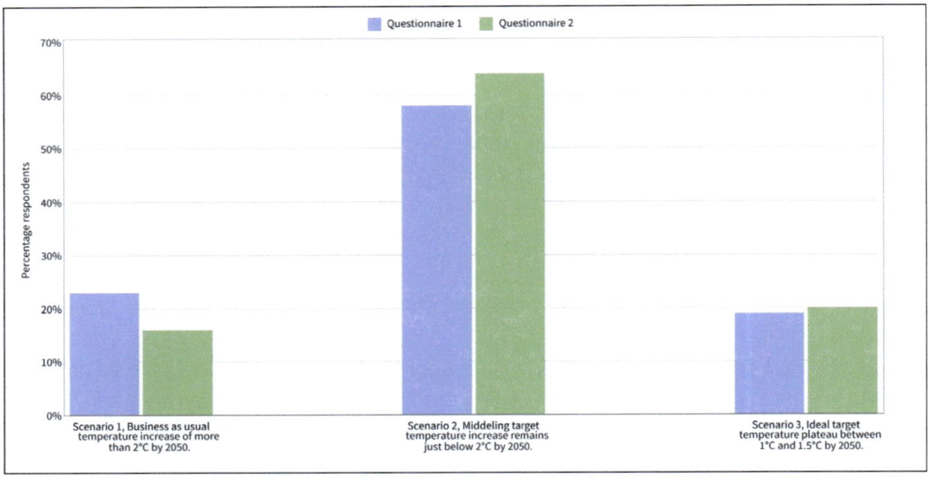

Source: Own compilation

2050; and scenario three, where global temperatures plateau at 1.5°C by 2050. This question was asked in the first questionnaire (n=50) (Appendix D, Table 7, p. 234), and repeated in the second (n=25) (Appendix D, Table 8, p. 234) to determine if there were any changes in the respondents' opinions over the six-month period (Figure 6.3).

In both questionnaires respondents overwhelmingly indicated that they thought scenario two, where the temperature remains just below 2°C by 2050 was more likely to happen (questionnaire 1: 58%; questionnaire 2: 64%). In the second round of questions, there was however a shift in the respondents' opinion, with the second most likely scenario now being scenario three instead of scenario one. However, in both questionnaires, the difference between the number of respondents who believed that either scenario one or three is most likely to happen, is marginal (in the first questionnaire the difference was three respondents, and in the second questionnaire only one). Overall, the responses are comparatively similar, even though the second questionnaire had half the number of responses of the first one.

6.6 Resource system (RS)

The resource system of the UNESCO AfriMAB network was established by investigating: the biosphere reserves represented in the questionnaire; those that had climate change projects; how many projects they started since 2000; the zones in which these projects were concentrated; and which biosphere reserves have conducted climate change impact, and CO_2 emission studies.

6.6.1 Which biosphere reserves do you represent or have knowledge of?

This question was asked in both the first and second questionnaires. In the first round, 44 respondents representing 34 UNESCO Biosphere Reserves (n=34), located in 20 countries, participated (Appendix D, Table 9, p. 235).

In many cases, some individuals represented several biosphere reserves within one country. South Africa had the most questionnaire respondents (13) who represented 8 biosphere reserves in total. The country with the second most respondents (9) was Côte d'Ivoire with five participants from Taï and five from Comoé Biosphere Reserves; one respondent represented both and therefore the number of representatives is only 9. In the second questionnaire (Appendix D, Table 10, p. 237), there were 33 participants who represented 35 biosphere reserves in 16 countries (n=35). Representatives of four countries chose not to participate in the second round, but this time, Ethiopia was added. This was possible because a respondent who previously represented biosphere reserves of another country, now represented Ethiopia. In total, representatives of only 26 biosphere reserves responded to both questionnaires (Figure 6.4).

In the second questionnaire, eight of the original biosphere reserves were no longer represented. However, in an unexpected shift nine new biospheres were represented. This is due to the participants' job mobility within the AfriMAB network: they may be representatives of companies closely affiliated biosphere reserves, researchers, or national authorities with changing involvement in biosphere reserves. Like with the first questionnaire, South Africa had the most questionnaire respondents (7), followed by Côte d'Ivoire (5) and Burkina Faso (4). This again hints that the less vulnerable a country is to climate change, the more representation it seems to have in conservation-minded institutions (Figure 6.4).

Figure 6.4 The 34 (37%) and 35 (38%) AfriMAB Biosphere Reserves were represented in questionnaire rounds one and two, where only 26 (28%) participated in both.

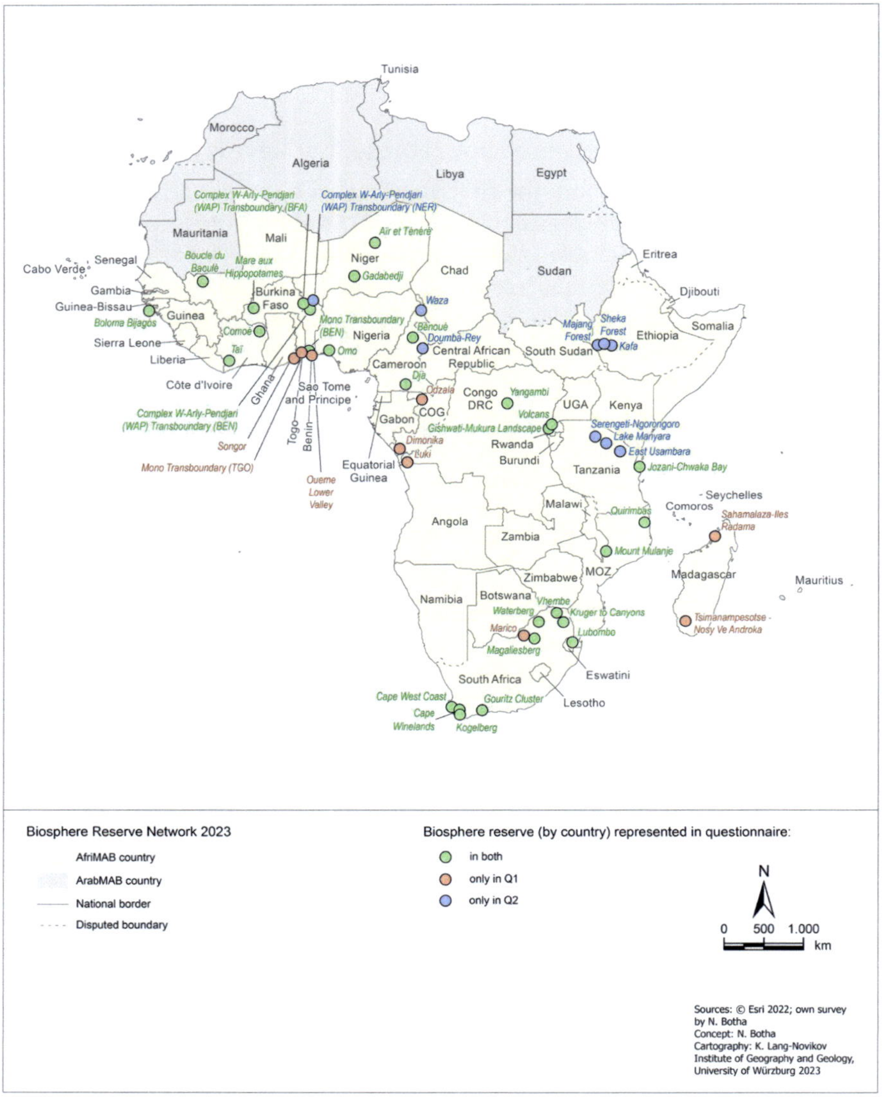

Source: Own compilation

Figure 6.5 Climate change projects across AfriMAB.

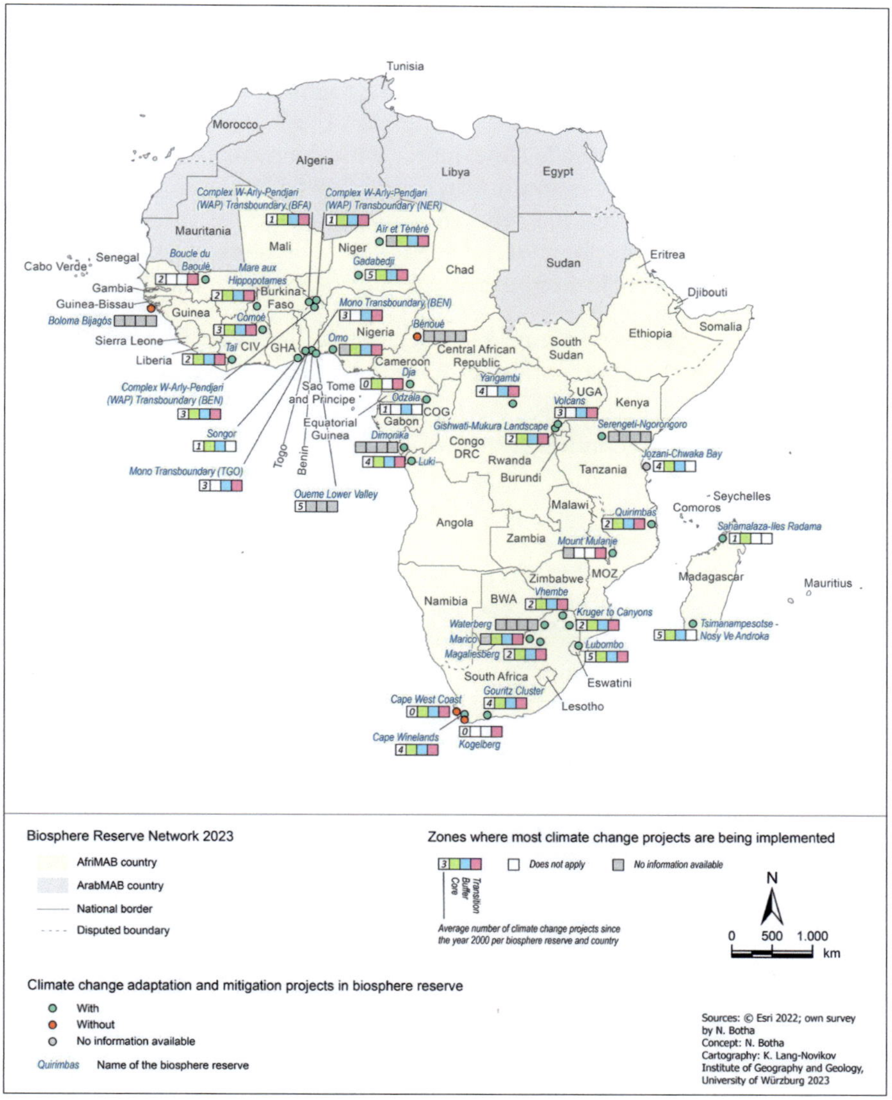

Source: Own compilation

6.6.2 In or near which UNESCO Biosphere Reserve(s) are climate change related projects of your organisation located? (n=50)

Respondents reported a total of 34 AfriMAB UNESCO Biosphere Reserves that have or had climate change projects (Appendix D, Table 11, p. 240) (Figure 6.5). Four of the biosphere reserves that participated in the first round of questionnaires did not have any climate change adaptation or mitigation projects, and three new biosphere reserves were mentioned. The country with the most biosphere reserves with climate change projects is South Africa (6) followed by Benin (3). The distribution of biosphere reserves with climate change projects is therefore spread across the AfriMAB network, however, there seems to be more of a concentration of these biosphere reserves in southern and western Africa.

6.6.3 Since 2000, how many climate change related projects have been pursued by your organisation within the biosphere reserve network? (n=27)

On average, respondents reported three projects per biosphere reserve (Appendix D, Table 12, p. 242). South Africa reported the highest number (14) of climate change related projects within its national network since 2000, followed by Benin (11), the DRC (8), and Eswatini (5). The Dja Biosphere Reserve in Cameroon showed an anomaly, namely that it reported having a climate change project, but also stated that it did not have one since 2000.

The biosphere reserves that ran the most climate change related projects since 2000 are Gadabedji (est. 2017) in Niger, Oueme Lower Valley (est. 2020) in Benin, Tsimanampesotse – Nosy Ve Androka in Madagascar (est. 2018), and Lubombo (est. 2019) in Eswatini, which each managed five (5) of these projects. The biosphere reserves with no reported climate change related projects were: Benoué (est. 1981) in Cameroon, Boloma Bijagós (est. 1996) in Cameroon, Kogelberg (est. 1998), and Cape West Coast (est. 2000) in South Africa.

6.6.4 In which biosphere reserve zone(s) are climate change adaptation and mitigation projects being implemented? (n=30)

The resource system within each biosphere reserve were divided into the core, buffer, and transition zones. Climate change related projects were reported to take place almost equally in each zone (n=30): the transition (83%) and/or buffer (83%) and the core (77%) (Appendix D, Table 13, p. 244) (Figure 6.5). The lower number of projects in the core zones can be ascribed to the fact that such projects need human involvement to help mitigate or adapt to climate change, and the core zones by design have fewer people.

6.6.5 Has there been research conducted on the impact of climate change on the UNESCO Biosphere Reserve(s) you are associated with? (n=20)

The results indicate that only 17 biosphere reserves in 14 countries conducted climate change impact studies (Appendix D, Table 14, p. 246) (Figure 6.5). That represents 43% (n=35) of the biosphere reserves that participated in the study. The three countries with the most (two each) UNESCO Biosphere Reserves that have done such an assessment are Cameroon, in Benoué and Waza; Côte d'Ivoire, in Comoé and Taï; and South Africa, in Kruger to Canyons and Vhembe.

Considering that biosphere reserves' fundamental function is logistic support through research (UNESCO, 1995b), it is surprising that less than half of them have done a climate change impact assessment. Such assessments may be seen as costly, but wasting resources on ineffective activities is even more so. For biosphere reserves to be effective tools for addressing climate change, actions should be based on information gathered through research, which is also the guidelines suggested by MAB (2011) and UNESCO (2016). Basing climate change projects on verifiable information significantly improves the chances of its long-term success. Research and learning is one of the fundamental principles of good governance, according to Folke et al. (2016).

6.6.6 Has there been research conducted on the amount of carbon dioxide (CO_2) emitted from your biosphere reserve? (n=30)

The results of the questionnaires show that only nine (27%) of the participating biosphere reserves conducted research on their CO_2 emissions (Appendix D, Table 15, p. 247). The two countries with the highest number of biosphere reserves that have conducted CO_2 assessments are Côte d'Ivoire, in Comoé and Taï, and the DRC, in Luki and Yangambi. This indicates an even worse performance than for research done on the impact of climate change. A reason for this may be that many of the countries belonging to AfriMAB do not emit much CO_2 compared to other regions, and that they therefore do not deem such research a priority. The results of such studies may however help these biosphere reserves to develop their (green) economies by identifying the sectors that are currently the biggest polluters, and to predict which sectors are likely to become bigger polluters in the future.

6.7 Resource system: Performance measures

In this study three resource performance measures were used, namely: natural environment resilience to a 2°C temperature increase (resilience); landscape connectivity (connectivity); and the degree to which the UNESCO Biosphere Reserve zones are functioning as mandated by the UNESCO MAB guidelines (adaptive capacity potential) (n=34). The average resource system performance was 60%

(Appendix D, Table 16, p. 248). Most (38%) of the respondents rated their biosphere reserves' natural environments' resilience to a 2°C temperature increase to be medium-low; their landscape connectivity was rated a medium (44%); and agreed (53%) that the biosphere reserves' zones are functioning as mandated (adaptive capacity potential).

The low majority percentages and average ratings indicate that respondents have not reached a consensus about the AfriMAB resource systems' overall performance, and that there is still room for improvement when it comes to managing the network. In terms of resilience, biosphere reserves should strive to cultivate a variety of strategies and obtain funding to address the potential impact of a 2°C increase, as suggested by Folke et al. (2016). Biosphere reserves should further aim to increase landscape connectivity, as it is vital to achieve the 30X30 goal and improve the overall environmental resilience. Of these indicators, adaptive capacity potential performed the best, and that means that that the post-Seville biosphere reserve model is in place in most cases and working as designed. Based on these responses to the questionnaires, it seems that in general, the AfriMAB network's resource system is located between the release and reorganisation phase on the adaptive cycle (Figure 6.6).

This phase is known for its formal chaotic behaviour with few resources, high diversity, and generally loose connections, with poor control over external variables (Holling and Gunderson (2002). At this stage, there is a greater possibility of foreign influences that will impact future growth. In terms of the resource system, it is now the perfect time to act to make a lasting difference. Figure 6.6 is a simplified visualisation of a more complicated and abstract concept; it is an illustration of the possible interactions within the resource systems, and of what may generally be happening in the network. Each biosphere reserve still has its own adaptive cycle and position on it.

In terms of individual performance scores, the biosphere reserves with the best performing resource systems are two of the oldest in AfriMAB, namely Yangambi (est. 1976) in the DRC, and Omo (est.1977) in Nigeria. These biosphere reserves also had most resource system indicators in place. The two biosphere reserves with the worst performing resource systems were of the youngest in the network, namely the Gishwati-Mukura Landscape (est. 2020) in Rwanda, and Lubombo (est. 2019) in Eswatini. Both had resource system elements in place, but according to their self-assessment, they were not performing well.

Figure 6.6 Resource system performance illustrated on an adaptive cycle.

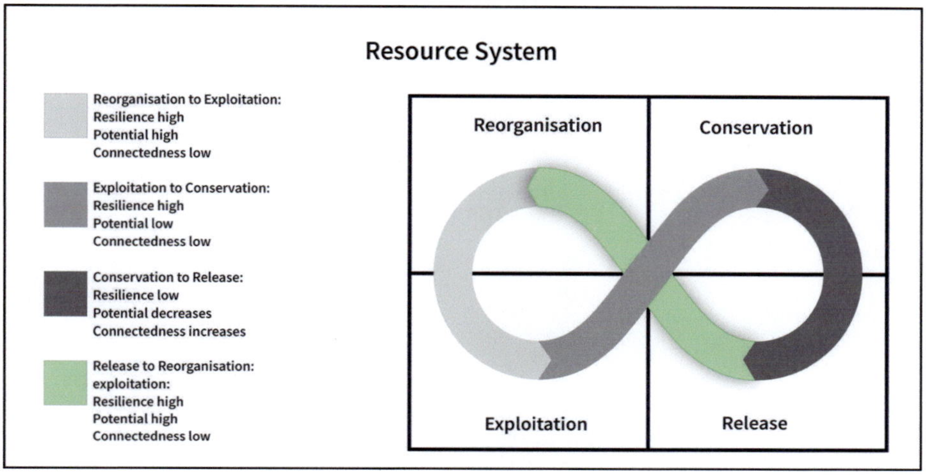

Source: Own compilation

These findings indicate a positive correlation between the age of the biosphere reserve and how well its resource system was performing. This may be proof that the UNESCO Biosphere Reserve designation positively impacts the functionality of the resource system in the long-term, and that as such it may already be contributing to the 30X30 target.

6.8 Resource units (RU)

This section investigated AfriMAB's current climate change adaptation and mitigation projects: project categories; when the implementation of the projects commenced; the duration of the projects; area coverage of projects; and the average annual cost of the projects.

6.8.1 Please name the climate change related projects you are currently busy with or have executed in the past? (n=26)

In total, 74 project categories were associated with 26 biosphere reserves, in 18 countries (Figure 6.7). The projects were divided into eight categories, namely:

Figure 6.7 Climate change project categories related to AfriMAB Biosphere Reserves.

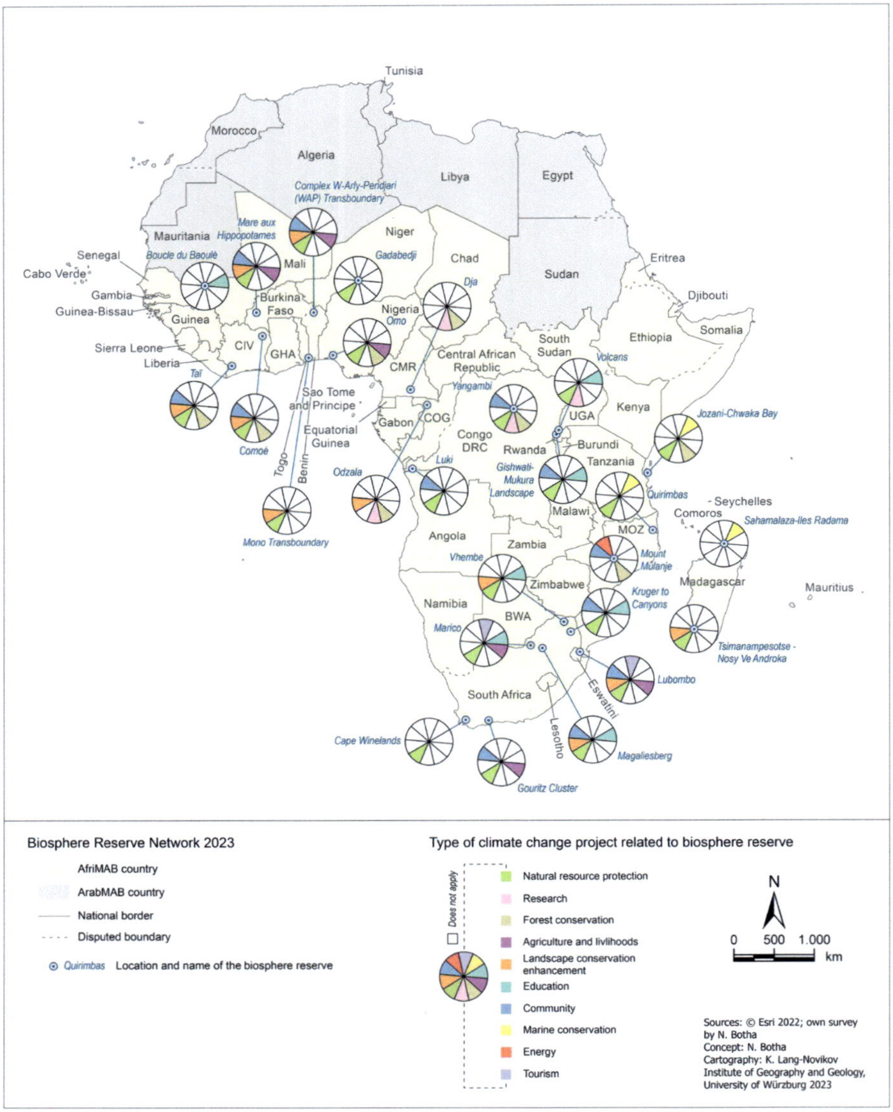

Source: Own compilation

natural resource protection (21), community (12), landscape scale conservation (10), forest conservation (8), education (7), agriculture and livelihood (6), research (4), marine conservation (3), tourism (2), and energy (1). Some of the projects may belong to more than one category.

Natural resource protection: natural resources are necessary to help mitigate and adapt to climate change, and these projects are run throughout the AfriMAB network (Figure 6.7). Respondents mentioned 21 climate change related projects dealing with natural resource protection, five projects addressing alien vegetation control, five on carbon sequestration, and three on the protection and management of water sources.

These projects included:

- Control of invasive alien vegetation
- Species protection
- Fire control
- Restoration and management of wetlands
- Restoration of semi-desert shrubland
- Protection and management of water sources
- Addressing water-related hazards
- Endangered hippopotamus project
- Landscape restoration
- Carbon sequestration
- Natural ecosystem management

Community: to be able to effectively adapt to and mitigate climate change, communities should be considered and encouraged to participate in projects. Twelve such projects were identified in AfriMAB (Figure 6.7). The goal of two of these projects were to enhance the resilience of the community. These projects related to:

- Climate change adaptation
- Resilience enhancement for climate change
- Community and school garden, and green programme
- Impact assessment of climate change on rural communities
- Community clean-up initiative
- Empowering people and nature
- Community-controlled grazing

Landscape scale conservation enhancement: this is needed to increase protection in conservation areas, and to expand their principles into the landscape.

This will aid ecosystem services in the mitigation of and adaptation to climate change. Ten projects aiming to enhance and enlarge conservation areas on a landscape scale, were reported. The projects were run mostly in the western and southern parts of AfriMAB (Figure 6.7), and related to:

- Ecosystem connectivity
- Expansion
- Protecting parks
- Biosphere reserve establishment
- Increasing protected area declarations
- Ecological corridor establishment
- Formation of informal protected areas

Forest conservation: Eight projects involving forest conservation were reported; the projects were run in countries around the equator (Figure 6.7). Three projects involved more efficient firewood stoves with lower fuel consumption; two projects involved tree planting with local community groups; and two projects related to community forest management. Four projects were focussed on the natural aspect of forest conservation for climate change mitigation and adaptation, while all eight also considered the social aspect of forest conservation.

- Protecting gallery forests and forest islands for regional climatic regulation
- Tree planting with community groups
- Community forest management
- Forest restoration
- Rehabilitate tracks and improve forests' fire brigades
- Forest restoration
- Low fuel consumption wood stoves
- Stop deforestation by providing an alternative livelihood

Education: education and the creation of awareness of climate change in communities are vital. Seven projects in five categories were mentioned by respondents; four of which focused on primary school children. These projects were mainly run in biosphere reserves located in southern and eastern Africa (Figure 6.7).

- Environment and climate change education in schools
- Weather monitoring school-to-school project
- Environmental education activities such as camps

- Improving the resilience of the educational system
- Raising general awareness of climate change

Agriculture and livelihood: these projects are important, as they focus on the prevention of deforestation and other activities degrading landscapes and ecosystem services vital for regulating and mitigating climate change. Three of the six overarching projects that were mentioned by respondents focused on conservation farming; and three were linked to the promotion of beekeeping, with one specifically focussing on the empowerment of women. These projects were concentrated in western and southern Africa (Figure 6.7):

- Development of a low-carbon local economy to conserve mangroves
- Domestication of wild fruit trees and introduction of useful plants into agroforestry
- Support to productivity and food security
- Stabilising livelihoods of vulnerable communities affected by conflict
- Biodiversity-friendly livelihoods
- Conservation farming
- Planting fruit trees at homestead
- Agroecology start-ups
- Beekeeping

Research: Four overarching climate change research projects with thirteen research topics were reported: two involved the monitoring of the long-term impact of climate change; eleven dealt with the natural science aspect and impact of climate change; four studies were on the social science aspect of climate change; and one on how research can be converted into policy. Most of these studies were conducted in the central AfriMAB region and involved forest conservation (Figure 6.7).

- Drylands and policy making
- Tree canopy changes
- Integrating native trees into forestry
- Fact-finding climate change issues in communities
- Building community resilience through an ecosystem-based approach
- Artificial intelligence for disaster risk
- Developing a digital reference for medical plants
- Biodiversity and climate science centre
- People and forest science centre
- Research addressing smallholder farmers' adaptation to climate change

- Research the interaction of biodiversity with climate change
- Monitoring long-term impacts of climate change
- Carbon credit assessment

Marine conservation: is important because climate change threatens ocean ecosystems. Three marine conservation projects were reported in eastern African countries (Figure 6.7), of which two aimed to enhance sustainable coastal fishing.

- Temporary octopus' hatcheries
- Crustacean fattening programme
- Marine sanctuary creation programme
- Sustainable coastal fishing
- Fish farming

Tourism: is a major component of African economies and their greenhouse gas emissions. Two tourism-related projects were reported, both in southern Africa (Figure 6.7).

- Training of tourist guides
- Tourism forums that encourage eco-tourism / low-impact tourism

Energy: the promotion of green energy and its effective use is key to greenhouse gas reduction. One energy-related project aimed at the mitigation of climate change was reported in Malawi, eastern Africa (Figure 6.7).

- Two hydropower plants
- Distribution of efficient electrical stoves

The climate change projects in AfriMAB mainly focus on the protection and enhancement of the natural landscapes and its ecosystems. It is therefore no wonder that the performance of the landscape's resource systems seems to improve with the age of the biosphere reserve. However, this is only one function of the biosphere reserves. Projects on the sustainable development function (agriculture and livelihood) were mentioned by less than a quarter (23%) of the representatives; these projects were mostly run in western and southern Africa. This is thus an area that still needs to be improved, especially since the previous results indicate that the more economically developed a landscape/region/country is, the better nature conservation representation is. This is why there were world-wide calls on the biosphere network in 1995, 2000 and again in 2014, to improve this function (UNESCO, 1995b, 2000, 2014).

Another area that needs improvement is once again research. Only 15% of biosphere reserves actively pursued research, but more importantly, most of this research is in field of natural sciences, while few studies involve the human aspect.

Considering that research is one of the functions of the biosphere reserves, the network-wide lack of research is surprising. In essence, the same problems concerning sustainable development and (social) research that were identified in the 1993 UNESCO internal report persist within the AfriMAB network 30 years later. The one aspect that has improved, is community involvement and education.

6.8.2 Do these projects relate to any of the following categories? (n=29)

On average, ten project categories were reported for each biosphere reserve (Appendix D, Table 17, p. 250). The categories mostly relate to the mitigation (62%) and adaptation (55%) of climate change in forestry, followed by mainstreaming biodiversity[3] (52%). This question uncovered a larger emphasis on sustainable development than the previous question. The three categories that were each only reported by one biosphere reserve were: low-carbon macro-economic approaches (regional marketing and circular economic flows) by Kruger to Canyons in South Africa; and the promotion of non-conflictual biofuels (e.g., agricultural residues or grass cuttings) in Lubombo, Eswatini; and adaptation to shifting vegetation by Cape West Coast, South Africa. Forestry mitigation and adaptation is therefore the main focus of biosphere reserve projects, which confirms the results of Möller (2011), however the economic considerations of climate change through mainstreaming biodiversity is much higher than was previously reported.

The three UNESCO Biosphere Reserves with the widest range of climate change project categories were Lubombo in Eswatini (66% of reported categories); followed by WAP transboundary in Benin, Burkina Faso and Niger (64%); and the Aïr et Ténéré Biosphere Reserve in Niger (53%). There was one biosphere reserve that reported a single project category namely, Gishwati-Mukura Landscape in Rwanda with its degraded (semi-) arid ecosystems restoration project. These results indicate that although the focus is on climate change projects related to nature conservation, there still is a range of objectives relating to various social and environmental climate change goals within each of the projects. According to Kotschy et al. (2015) this diversity increases the probability that a climate change project will be successful in the face of adversity – which is prevalent in the AfriMAB network.

6.8.3 When did UNESCO Biosphere Reserves become involved with climate change related projects? (n=23)

The DRC's Luki and Yangambi Biosphere Reserves reported the earliest (1979) two climate change projects (Appendix D, Table 18, p. 254). The third oldest climate

3 Mainstreaming biodiversity is when biodiversity is directly incorporated into the policies and planning of business or industry to encourage sustainable development (OECD, 2018).

Figure 6.8 Year the first climate change related projects were implemented.

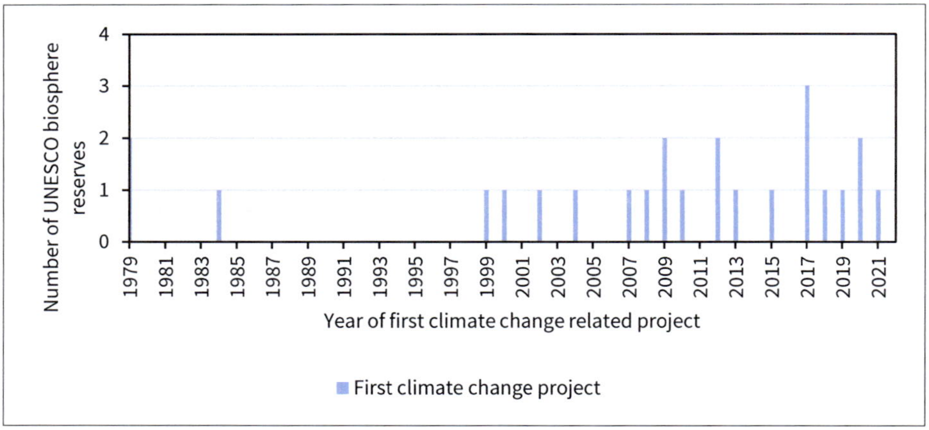

Source: Own compilation

change project was initiated in 1984 in the Dja Biosphere Reserve, Cameroon. The biosphere reserve that has most recently (2021) started to run climate change related projects, is Cape Winelands in South Africa.

After the Kyoto Protocol was signed in 1997 (von Weizsäcker et al., 1999), the number of climate change projects started to increase (Figure 6.8). After 2009 a marked number of biosphere reserves initiated their first climate change projects (Figure 6.8). In 2007 the 3rd World Congress of Biosphere Reserves was held in Madrid, after which the first biosphere reserve action plan was published (UNESCO, 2008). Around 2012 there was again an increase in the number of projects (Figure 6.8), after the 40th Anniversary Conference of the MAB programme in Dresden in 2011 (MAB, 2011).

Other events during this period were the 2011 CBD Aichi target announcement, and the publication of the UN sustainable development goals in 2012 (CBD, 2011; Sachs, 2012). After the 2016 publication of the Lima Action Plan, there was an uptick in the number of climate change projects in 2017 (Figure 6.8), indicating that biosphere reserves seem to be responsive to UN and UNESCO Biosphere Reserve events, and that they will likely be inspired by them to take climate change action. The results also show that there is often a delay of about a year between the events and the start of a new project, as members probably need some time to plan a project and obtain funding to execute it. The momentum of these first-time climate change projects on average seems to last 4 years (Figure 6.8). This indicates that the World Network of Biosphere Reserves' meetings that set the biosphere reserves' agendas for each decade, should ideally be held every five years instead of every ten years in order to encourage and maintain enthusiasm for climate change actions.

6.8.4 What is generally the intended area size of climate change mitigation and adaption project in your biosphere reserves? (n=23)

Most biosphere reserves reported that the average intended size of their projects is between 51-60% of the biosphere reserve (Appendix D, Table 19, p. 255). Two reported an aim of more than 100% of the biosphere reserve; this is likely where the projects are managed by external organisations. The biosphere reserve with the largest intended project area is Aïr et Ténéré in Niger, encompassing 95,880 km^2 or 31-40% of the biosphere reserve. The second largest project area is the Complex W-Arly-Penjari (WAP) Transboundary, covering 53,619.90 km^2 or 51-60% of the biosphere reserve. Vhembe in South Africa is the third largest with 24,080.80 km^2 or 71-80%. The smallest intended project area is Mount Mulanje in Malawi with 25.68 km^2, covering 31-40% of the biosphere reserve. This is followed by the Magaliesberg with 71.57 km^2 (1-10%) and Kogelberg 103.62 km^2 (1-10%) biosphere reserves in South Africa.

These findings indicate that most biosphere reserves intend to run climate change mitigation or adaptation projects on around half of their land. In an ideal world, when this figure is extrapolated to the entire AfriMAB, climate change adaptation and mitigation projects would cover between 513,793 km^2 (50%) and 616,551(60%) km^2. In terms of entire sub-Saharan Africa this figure translates to between 2.1% and 2.5% of the region.

6.8.5 On average, what is the duration of climate change mitigation or adaptation projects in your biosphere reserve(s)? (n=26)

Most (35%) biosphere reserves reported that climate change projects are expected to last about four years (Appendix D, Table 20, p. 257). This was followed by 15% of biosphere reserves reporting that climate change projects only last on average one year. The biosphere reserve that reported the longest average duration of climate change related projects was Mount Mulanje in Malawi with more than twenty years. This was followed by Yangambi in the DRC, and Dja in Cameroon, with an average project duration of between ten and 19 years.

Western Africa's projects consistently last around four years, and the central region's projects last much longer than most in AfriMAB (Figure 6.9). Eastern and southern Africa's projects are a mix of short- and long-term projects. The duration of the projects is important as the longer the project last, the more impactful they are likely to be. This may ensure that climate change related projects will be consistently run on between 2.1% and 2.5% of the region.

6.8.6 Provide an estimate of the average annual cost per climate change project. (n=16)

The highest estimated annual cost of a climate change project was USD 15,509,577.00 in Waterberg, and the second highest was USD 1,072,495.00 in the Gouritz Cluster; both biosphere reserves are located in South Africa (Appendix D, Table 21, p. 258). Western Africa had the lowest estimated annual cost per climate change project in the AfriMAB network, central Africa on average had a higher financial input, while climate change projects in southern and eastern Africa had a mix of higher and lower estimated annual costs (Figure 6.9). This most likely relates to the type and origin of the funding that these biosphere reserves are able to obtain for their climate change projects.

These estimates relate to specific types of projects; while Waterberg did not indicate the type of projects that they fund, the Gouritz Cluster reported that their projects address natural resource protection, community, as well as agriculture and livelihood enhancement. The third highest estimate was USD 1,057,082.45 by a national representative from Tanzania, for East Usambara, Jozani-Chwaka Bay, Lake Manyara, and Serengeti-Ngorongoro. The only biosphere reserve from Tanzania that answered the question regarding the types of projects, was Jozani-Chwaka Bay; stating that their projects addressed marine and forest conservation, and natural resource protection.

The lowest estimated annual cost of climate change related projects was reported by Mare aux Hippopotames in Burkina Faso, with USD 245 for projects relating to natural resource protection, agriculture and livelihood, community, and landscape conservation enhancement. On average, when the unusually high estimate provided by Waterberg is included, climate change related projects annually cost biosphere reserves about USD 1,379,790.24. However, when calculating the average estimate without the Waterberg outlier, the average estimate decreases to USD 437,804.45 annually.

The duration of climate change projects, the intended area covered, and the annual cost are interesting aspects to consider together (Appendix D, Table 22, p. 259). Unfortunately, only representatives from seven biosphere reserves answered all three questions (n=7). While the representative from Tanzania answered all the questions, it is unclear whether they answered on behalf of the network or the individual biosphere reserve(s), and therefore their reply had to be left out of this equation. Kogelberg in South Africa was also omitted as they reported that they had no climate change related projects, so their estimates must therefore be hypothetical.

Figure 6.9 Average annual duration and cost of climate change projects in AfriMAB.

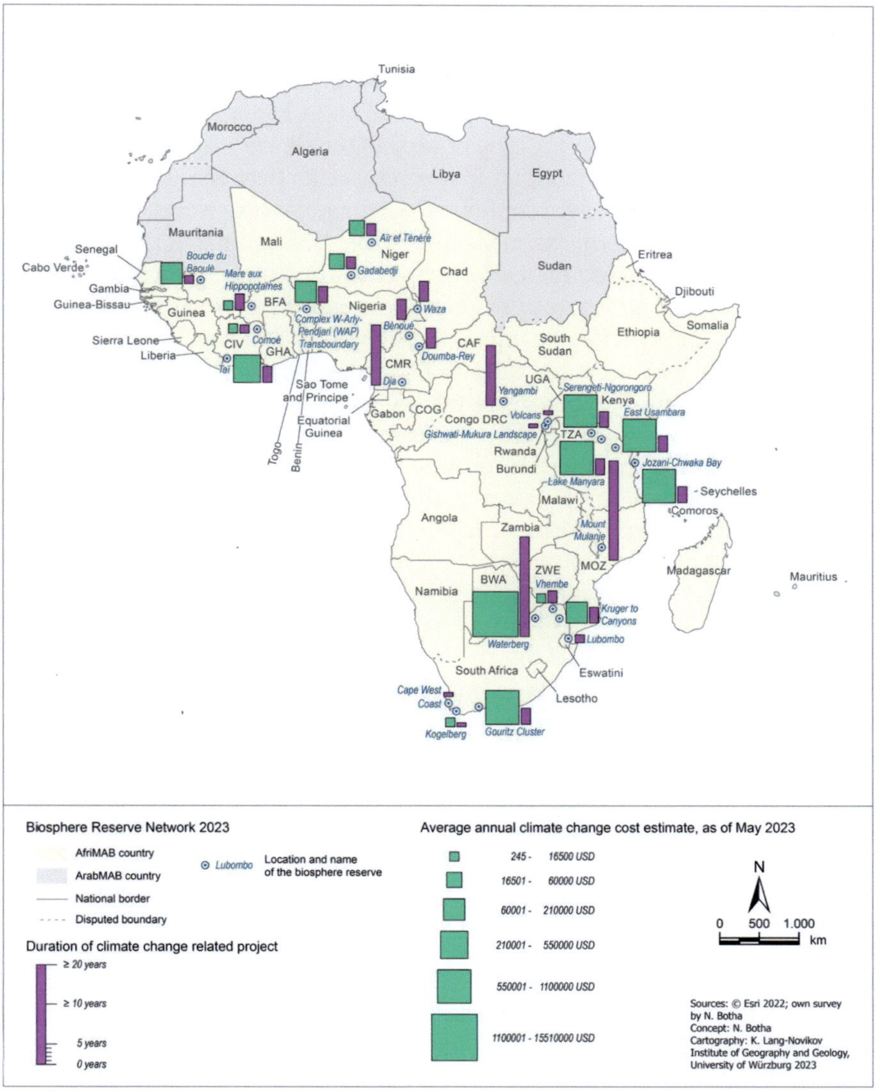

Source: Own compilation

The small remaining sample of useful responses indicate that the average estimated duration of climate change projects are three years, with an average estimated cost of USD 6.25 per km^2 per year. This cost seems to vary greatly from country to country, with the highest being USD 18.73 per km^2 for education project run by the Boucle du Baoulé Biosphere Reserve in Mali. The lowest cost for climate change related projects was USD 0.63 per km^2 in Aïr et Ténéré in Niger. This biosphere reserve did not disclose the type of project they were working on.

The results therefore indicate that it currently costs between USD 3.2 and 3.9 million (at USD 6.25 per km^2) per year for at least three years to run climate change adaptation and mitigation projects in 50-60% of the AfriMAB network, which represents 2.1% to 2.5% of the region. However, as the aim of this study is to improve the current institutional structure of the biosphere reserve network to advance the CBD's 30X30 target and specifically improve its environment's resilience to climate change, a more aggressive budget of USD 20 per km^2 is suggested. This number was chosen as it was rounded up from the highest reported funds per km^2. For climate change actions to succeed they need sufficient funds and the higher funds of USD 20 per km^2 could act as a starting point for biosphere reserves to budget for their climate change projects. This is why the collected information was extrapolated to USD per km^2. Based on this estimate, the AfriMAB network currently needs between USD 10 and 12 million per year for its climate change projects, but this amount must at least be doubled to fully cover every biosphere reserve in the network.

Because this estimate is based on such a small sample of biosphere reserves, it is only meant to be a guideline of what may be needed to ramp up climate change related projects in the AfriMAB network. Because all biosphere reserves are different, as their landscapes, demography and threats differ, their project costs are also different. A proper in-depth financial analysis of climate change projects in the AfriMAB network is needed to determine the exact requirements for each biosphere reserve.

6.9 Resource unit: Performance measures

In this section, we only use two performance measures. The first is to determine whether respondents have witnessed a marked positive impact of their climate change adaptation and mitigation projects (an indicator of its adaptive capacity potential) (n=31). The second is to consider how resilient climate change related projects have been to past global (disruptive) events (e.g., the Covid-19 pandemic) considering that the prevalence of such events are likely to increase in the future.

The average overall performance score of the resource units (climate change related projects) was 57% (Appendix D, Table 23, p. 260). Most biosphere reserves

Figure 6.10 Resource unit performance illustrated on an adaptive cycle.

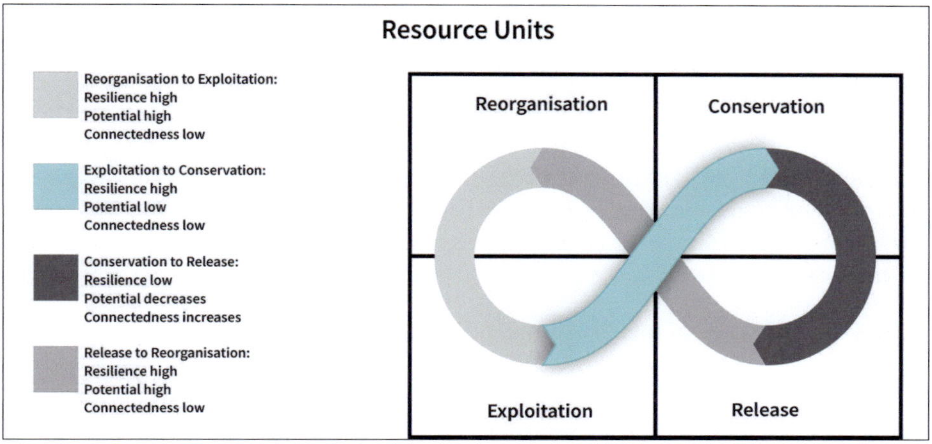

Source: Own compilation

reported that their climate change adaptation and mitigation projects "maybe" (35%) had a positive impact, and that these projects have a "medium" (35%) resilience, which shows that there is room for improvement. According to Folke et al. (2016) to do this, biosphere reserves must run a variety of climate change projects with multiple funding sources to cultivate resource unit resilience. To determine whether these projects have made a difference, research and monitoring should be encouraged. Nonetheless, based on most of the resource unit's performance ratings, the system is located between the exploitation and conservation phase of the adaptive cycle (Figure 6.10). The low number of resource unit performance votes – even if it is the majority of votes – infer that a consensus among the AfriMAB network was not reached.

According to Holling and Gunderson (2002) this phase is known for its increasing stability and predictability. While an unlimited number of opportunities is available, there is also an increased competition for resources; some projects may fail, but those that succeed become dominant in the landscape. It makes sense that the AfriMAB network finds itself in this phase of the adaptive cycle, as climate change projects have been increasingly encouraged, managed, and improved within the network since 2000. The results indicate that the two biosphere reserves with the best performing resource units are Boucle du Baoulé in Mali (100%) with its education project, and Mare aux Hippopotames in Burkina Faso (86%), with its natural resource protection, agriculture and livelihood, community, and landscape conservation projects. The three worst-performing biosphere reserves in terms of their resource units, with 14% each, are Aïr et Ténéré and Gadabedji in Niger, and Yangambi in the DRC.

6.10 Governance systems (GS)

The governance systems of the AfriMAB network were determined by examining the following aspects: the degree to which climate change was integrated into the biosphere reserves' institutional structure; the strategies adopted to promote climate change adaptation and mitigation; the role of the biosphere reserve as a conservation body in climate change related projects run in their area; and the source of funding for climate change projects.

6.10.1 The degree of integration of climate change action within the biosphere reserves' institutional structure. (n=30)

The degree of integration was established by investigating the extent to which the following elements were implemented: whether the biosphere reserve set up a budget for climate change related projects; the importance of climate change in the management plan; the presence of a formulated climate change strategy/action plan; and the degree to which the biosphere reserve positioned itself specifically in terms of climate change (Appendix D, Table 24, p. 262).

Most (70%) of the biosphere reserves indicated that they have "fully" included climate change as an important issue in their management plan, but a much lower number (53%) reported that they had in fact ("mostly") formulated a climate change strategy/action plan; or either "mostly" (37%) or "partly" (30%) positioned themselves specifically to address climate change. The element that performed the poorest was budgeting for climate change projects as most (47%) biosphere reserves reported that they only "partly" budget for climate change projects.

When comparing these results with an earlier study by Möller (2011) on the World Network of Biosphere Reserves, a marked increase in the implementation of all these elements is apparent. The most significant change was in making climate change an important issue in the management plan. This improvement was likely a result of the impetus provided by the World Network of Biosphere Reserve meeting in 2016.

Figure 6.11 Institutionalisation of climate change within management structure.

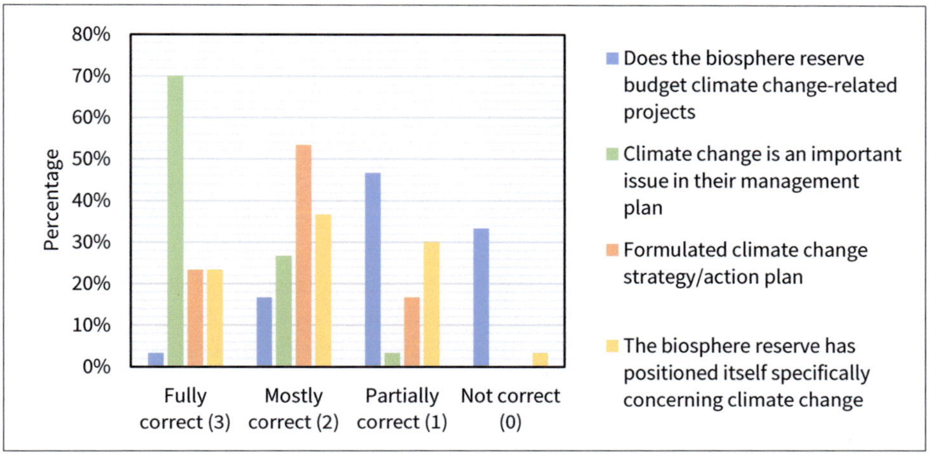

Source: Own compilation

The total institutionalisation scores show that the best-performing biosphere reserve is the Gouritz Cluster in South Africa (92%). Taï and Comoé in Côte d'Ivoire; Gishwati-Mukura Landscape and Volcans in Rwanda, and Sahamalaza - Iles Radama in Madagascar, all scored 83%. The average score was around 60%. The worst performing biosphere reserves were Tsimanampesotse - Nosy Ve Androka in Madagascar (25%), Kogelberg in South Africa (25%), Dja in Cameroon (33%), and Odzala in the Republic of the Congo (33%). The performance variation in the AfriMAB network (Figure 6.12) can be attributed to the vastly different contexts of the individual biosphere reserves, including the governance structures and their role in climate change projects.

6.10.2 What is your biosphere reserves' role in climate change projects? (n=28)

The majority (79%) of respondents reported that their UNESCO Biosphere Reserve had a partnership role, 68% said they had a leading role, 50% a coordinating role, 18% an observer role, and 4% reported that they had no role (Appendix D, Table 25, p. 265). The various regions in AfriMAB seem to have some clusters of climate change project roles. In western Africa there are a cluster of leading, partnership, and coordinating roles. In central Africa partnership roles are dominant, while in southern Africa a combination of leading and coordinating roles appear more frequently (Figure 6.12).

The three biosphere reserves that had the largest variety of roles in climate change projects were: WAP in Benin, Burkina Faso; Gouritz Cluster, and Vhembe Biosphere Reserves in South Africa. One biosphere reserve had reportedly no role in climate change projects, and this was Cape West Coast in South Africa. These differing levels of involvement in biosphere reserves are possible through partnerships with and funding from external organisations.

6.10.3 Are these projects funded on an international, national/federal, state/provincial, or local level? (n=28)

Representatives reported that biosphere reserves obtained most of their funding for climate change related projects from either international (93%) or local (61%) sources (Appendix D, Table 26, p. 267). The federal/national and state/provincial funding sources were each stated in only 32% of the questionnaires (Figure 6.13). The need for an emphasis on international cooperation and local involvement agrees with views published by MAB (2011), and Stoll-Kleemann and O`Riordan (2017).

The first questionnaire's preliminary analysis showed that 49% of all climate change projects' funding is sourced internationally, 20% locally, 12% nationally/federally, and 11% from the state or province. These numbers were based on n=53, which is the total number of respondents, and not n=24, the number of participating biosphere reserves; the statistics were skewed because multiple respondents gave answers about the same biosphere reserve.

Figure 6.12 AfriMAB's institutionalisation and role in climate change projects.

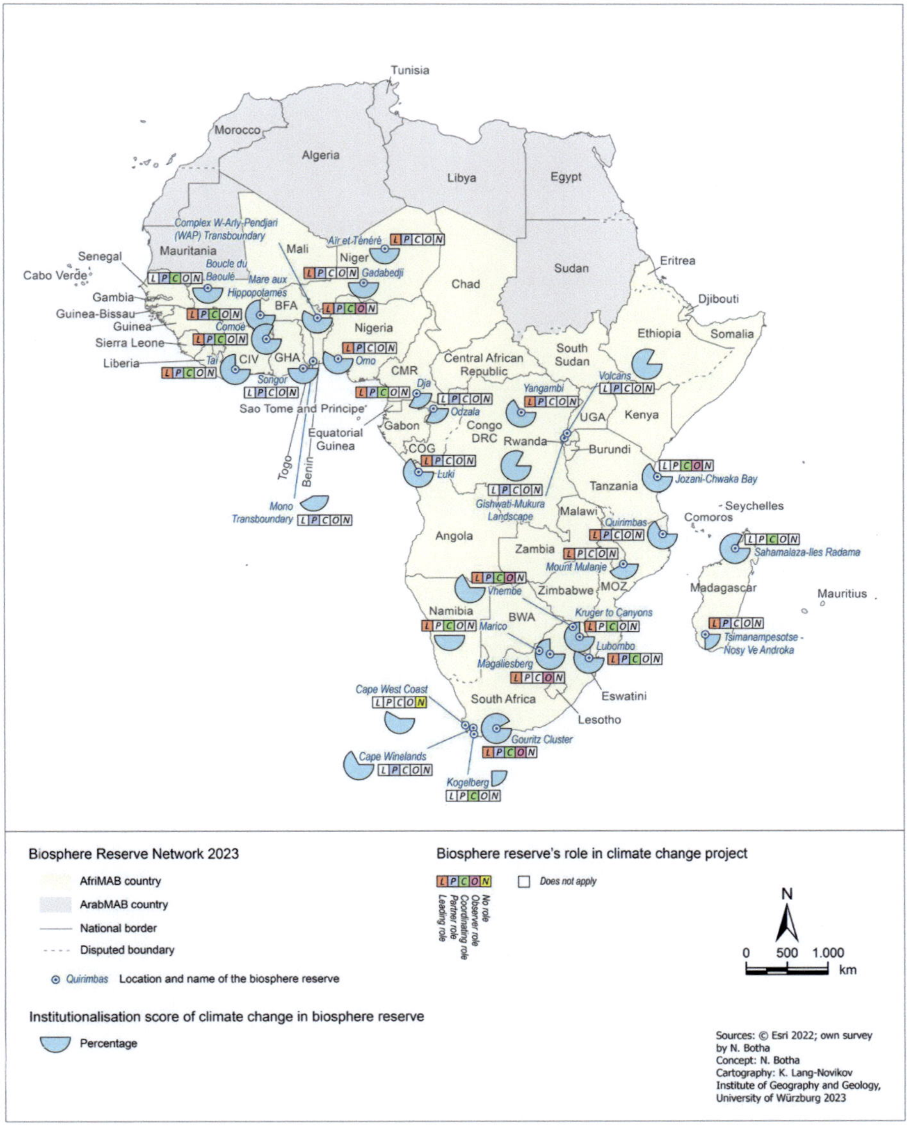

Source: Own compilation

Figure 6.13 Climate change project funding origin.

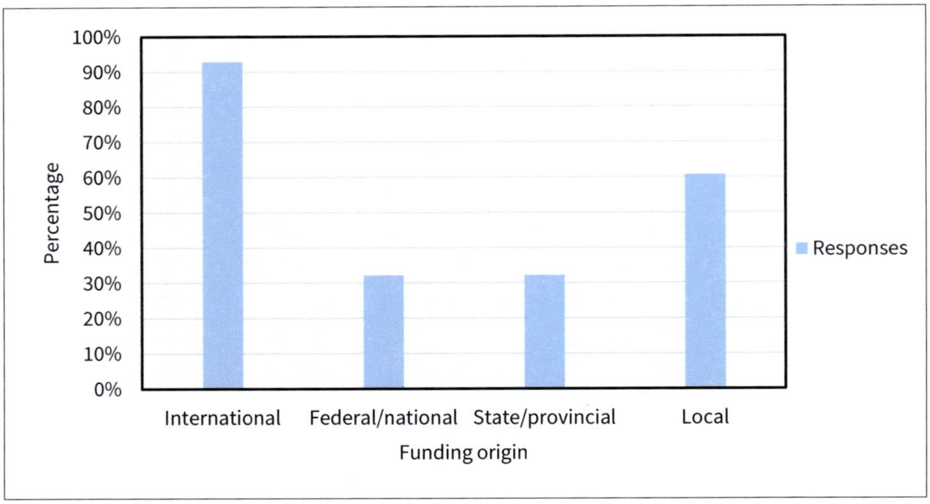

Source: Own compilation

This oversight was discovered with the completion of the second questionnaire, in which respondents were asked if they agreed with these percentages. Some apparently picked up on this mistake as 31% indicated that international funding should be higher, between 80-100% (n=28). Nevertheless, 46% believed that the original percentages were plausible. While the corrected percentages are now higher for each category, international funding is still the highest, followed by local funding, and federal/national and state/provincial almost equally the lowest.

Respondents were asked to elaborate on their answers in the second questionnaire (n=28). Some respondents believed that there is a lack of local funding, but participants from Rwanda and South Africa felt that local funding is higher because the impact of climate change is more prominent on a local scale, and that it is therefore important that climate change projects should be locally designed and implemented. Respondents from Burkina Faso and South Africa said their governments should contribute more, but that they do not take climate change seriously; these governments neglect the environmental sector and prioritises the social sector.

Although most of the respondents were national level stakeholders, the need to find funding when it is not available locally, makes sense. However, the limited federal/national and state/provincial funding for biosphere reserves suggests that governments across the AfriMAB network generally do not acknowledge their role in climate change strategies. This means that that the AfriMAB Biosphere Reserve network cannot perform optimally to mitigate and adapt to climate change in terms of the Lima Action Plan (UNESCO, 2016). Nothing has changed since UNESCO observed in 2010, and Reed confirmed in 2016, that biosphere reserves remain under-utilised by government agencies. Perhaps the UNESCO MAB programme should start a campaign to encourage governments to recognise the role of biosphere reserves in climate change mitigation and adaptation, like the IUCN did with their protected area categories.

6.10.4 At which UNESCO MAB level should climate change adaptation or mitigation initiatives originate for the project/initiative to be sustainable? (n=31)

Most (81%) respondents believed that climate change projects must originate locally, or at least nationally (32%) (Appendix D, Table 27, p. 268) to be sustainable. Only 10% said it must originate internationally, while 16% thought that the projects should originate elsewhere (Figure 6.14). Two respondents explained that each level has something to offer, while two other respondents emphasised the importance of the local sphere and highlighted that all climate change related projects should consider the expectations of the local community.

The local biosphere level is therefore extremely important to the success of climate change adaptation and mitigation projects. This supports the findings of Schliep and Stoll-Kleemann (2010), Cuong et al. (2017b), and Stoll-Kleemann and O`Riordan (2017).

Figure 6.14 Initiative's origin for the project/initiative to be sustainable.

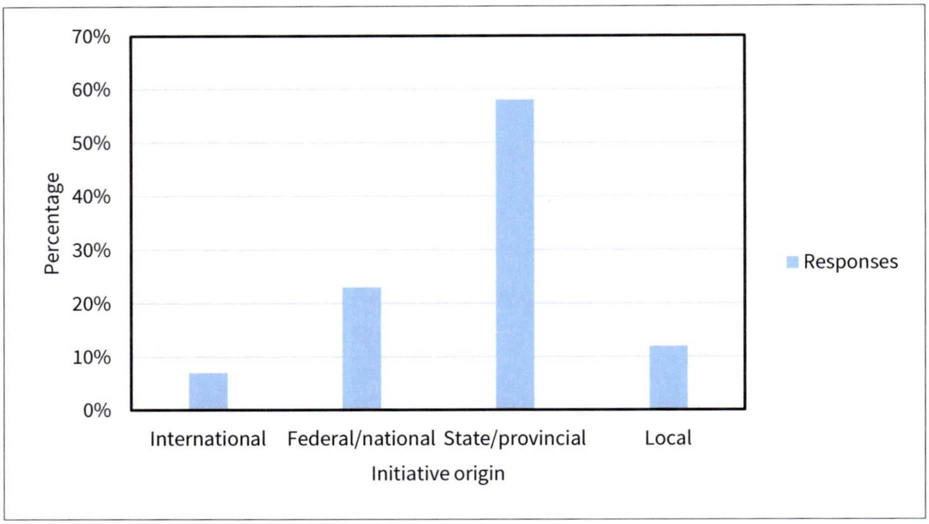

Source: Own compilation

6.10.5 Does the organisation you represent implement the following climate change strategies? (n=29)

Respondents indicated that their biosphere reserve participated in, on average, six of the twelve climate change strategies (Appendix D, Table 28, p. 269). The top strategies were the creation of public awareness/understanding (90%); involving new stakeholders, including traditional and indigenous communities from outside the biosphere reserve (83%); sensitizing decision-makers and donors about climate change (76%); and adopting climate change education for children and students (76%). The strategies that were mentioned the least were: involving political/economic stakeholders in research designs (34%), and adapting the planning, implementation, and evaluation processes of the management plan (34%). These findings agree with Möller (2011), especially regarding the creation of awareness as a key strategy. It suggests that the post-Seville biosphere reserve shift towards more inclusive community projects is in fact taking place. However, it also confirms that there is an issue with involving government stakeholders in climate change projects/initiatives.

6.11 Governance system: Performance measures

This study used the following four governance system performance measures: connectivity between the biosphere reserves and international, national and regional networks; the perceived resilience of its institutional structure; and also, as a collective measure of adaptive capacity potential, the extent to which the biosphere reserves address environmental reality and local needs, while considering the governing body's capacity for long term implementation, in their climate change projects (n=31) (Appendix D, Table 29, p. 271). The governance system scores indicate that the top performing biosphere reserves were Omo in Niger (100%), Boucle du Baoulé in Mali (88%); Comoé in Côte d'Ivoire (84%); and Gouritz Cluster (84%) and Vhembe (84%) in South Africa. The worst performing biosphere reserves were Kafa, Majang Forest, and Sheka Biosphere Reserves from Ethiopia, each scoring 36%.

The results indicate that the average governance system performance score is 65%. Most biosphere reserves had a "strong" relationship with both the international (29%) and regional (32%) networks, and a "strong-medium" relationship with the regional network (42%). Overall, the connectivity between the biosphere reserve networks and most of the levels of government can be classified as "strong". While most of the ratings are good, they all have rather low percentages. According to Simonsen et al. (2014) strong connectivity within a network means that the chances are better that important resources and services will be distributed throughout the network; it also aids fast response times during crises.

Most respondents (55%) classified the resilience of the biosphere reserves' institutional structure as "medium". Of the respondents, 48% agreed with the statement that biosphere reserves address both environmental realities and local needs, while still considering the governing body's capacity for long term implementation, in their climate change projects. The final scores indicate that the top performing biosphere reserves were Omo in Niger (100%), Boucle du Baoulé in Mali (88%); Comoé in Côte d'Ivoire (84%); and Gouritz Cluster (84%). and Vhembe (84%) in South Africa. The worst performing biosphere reserves were Kafa, Majang Forest, and Sheka Biosphere Reserves from Ethiopia, each scoring 36%.

Figure 6.15 Governance system performance illustrated on an adaptive cycle.

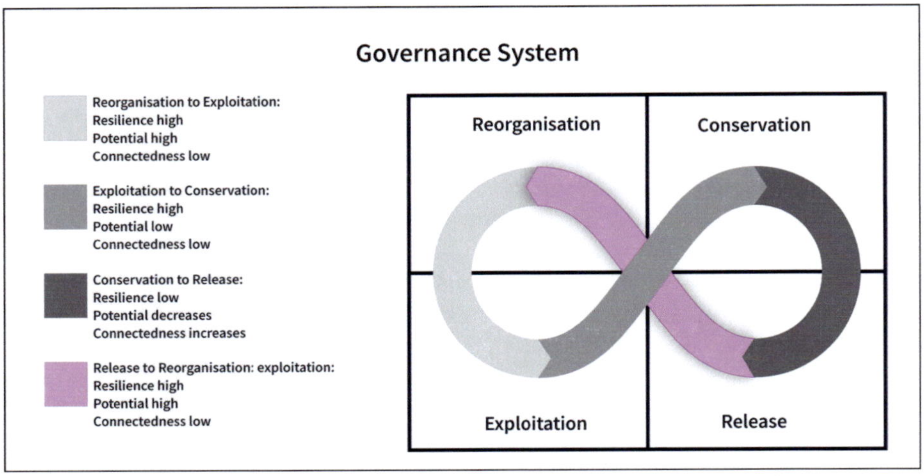

Source: Own compilation

The overall findings of this section of the questionnaire show that the governance systems of biosphere reserves is also located on the release to reorganise leg of the adaptive cycle (Figure 6.15), but considering the size of the majority representatives in each category, there is not a clear consensus in the network about the performance of the governance system.

This classification is reasonable considering the diversity of governing structures within AfriMAB, and the general scarcity of resources for projects. Considering the lack of participation in international and especially UN events, and the overall vulnerability of AfriMAB to climate change, control over external variables can definitely be classified as weak. Actions taken at this point will greatly influence the future of the governance system.

6.12 Actors

Actors (e.g., government officials, biosphere reserve board members, local communities, and business) are people with an active role in AfriMAB. Who the actors in the network are, was established by asking respondents about the group or groups that they represent within the network; to indicate which of those groups they thought should be responsible for climate change projects for it to be sustainable; which group they thought should initiate research for it to be actively implemented; which other individuals/organisations are implementing climate change projects within the biosphere reserve; how active are these organisations in the different zones of the biosphere reserve; and what their future role will be. During the first questionnaire respondents were also asked what their level of awareness of the impact of climate change was before their participation in the study.

6.12.1 Which stakeholder group(s) do you primarily represent in the UNESCO MAB Biosphere Reserve network? (n=66)

Each stakeholder was involved in at least two groups (Figure 6.16). The largest group (39%) was government organisations: of this group 62% of the respondents (n=26) represented the national government, 19% the local government, and 15% the provincial/regional government. Of the overall number of respondents (n=66), 33% served on the national biosphere reserve committee, and 33% acted as the biosphere reserve's coordinator or chairperson. The third largest group of respondents (26%) was from the formal conservation area management group. Only three respondents belonged to the thematic/regional network membership group.

Figure 6.16 Stakeholder groups to which respondents belong.

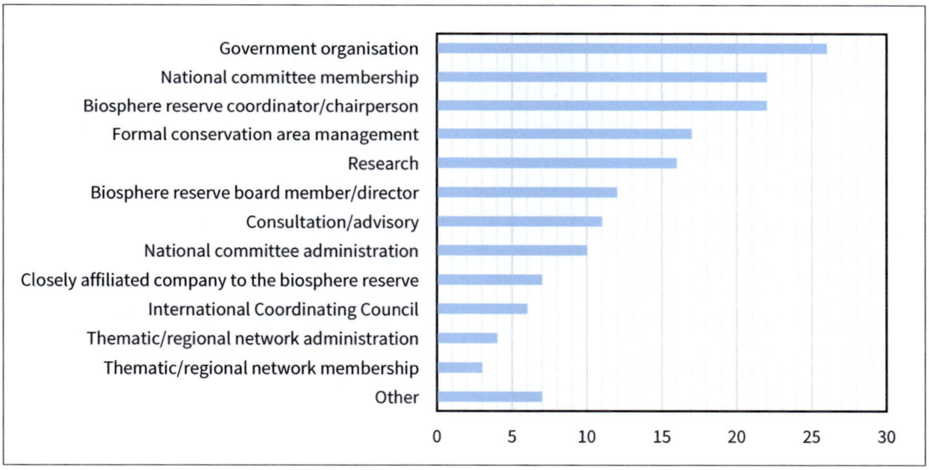

Source: Own compilation

Respondents associated with the following stakeholder groups were subsequently asked to elaborate on these categories: research, board members, consultation advisers, and closely affiliated companies. Of the researchers (24% of respondents), the biggest group (6% of the researchers) were involved in ecological and conservation research. The rest mostly indicated specific research interests, rather than a general field of study. These included; agronomy and soils; animal biodiversity, ethno-zoology, and dynamics; forest botany; natural sciences; sea turtles, small pelagic, ecosystems, and aquatic birds; biodiversity of flora and fauna; local knowledge on the use and conservation of natural resources; ecosystem services in biosphere reserves; tropical forest restoration, primate ecology, conservation, ecosystem-based adaptation, and buffer zones; and wetlands, micro and macro corridors, dryland soil erosion control, and habitat restoration. The researchers were from organisations such as the University of Rwanda, Rwanda; Nangui Abrogoua University, Côte d'Ivoire; National Centre for Scientific and Technological Research, Burkina Faso; and the University of Eswatini, Eswatini.

Several respondents (18%) had positions as biosphere reserve board members. Their portfolios included: chairpersons (4%), heads of divisions (3%), managers (2%), CEOs (2%), conservation directors (2%), marketing directors (2%), biosphere reserve task force members (2%), and unspecified (1%).

The consultation and advisory stakeholder group formed 17% of all respondents. Their fields of expertise included biodiversity; biodiversity and sustainability; evolutionary biology, wildlife biology, conservation biology, ecology, foraging biostatistics, agricultural economics, and energy; management and governance; nature conservation; ecosystem services use; biosphere reserve zonation; conservation management; and water and water-related issues.

Lastly, respondents from closely affiliated companies formed 9% of all stakeholders. They included representatives from non-profit organisations (2%); CIMTOGO, a cement company (2%); the Sustainable Management of Biodiversity and Protected Areas Project (SMBAP) (2%); Eswatini National Trust Commission (1%); Green Africa Group (1%); and the German Technical Cooperation (GIZ) (1%).

Thematic network administration (6%) and membership (5%) were the least represented stakeholder groups. Some respondents (13%) did not identify with any of the stakeholder categories and classified their involvement as "other". This category includes youth network members (3%); the African Parks Network (2%); a conservationist of a reserve (2%); a core employee of a biosphere reserve (2%); a member of a non-governmental organisation (2%); and a site manager (2%).

These results indicate that most of the stakeholders hold government, nature conservation or governance positions, therefore respondents should have been aware of COP15 and COP27, and more respondents should have had knowledge of what was discussed at these meetings. Also, the results indicate that respondents representing the research and advisory/consultant group almost exclusively represent the field of the natural sciences. This means that the post-Seville biosphere reserve model's new goals of expanding biosphere reserve research into the social sciences have not yet taken effect in AfriMAB.

6.12.2 Which stakeholder(s) should be responsible for climate change mitigation and adaptation projects for it to be sustainable? (n=33)

Respondents generally provided more than one answer (Appendix D, Table 30, p. 272), but most (72%) indicated that government organisations should be responsible; as well as the biosphere reserve coordinator/chairperson (70%); and national committee administration (61%). The two lowest-scoring groups were thematic/regional network members (12%) and people in consultation/advisory positions (15%). Based on these findings respondents agree that (national) governmental support is needed in conjunction with local execution to address climate change. This was proposed at the 40th Anniversary Conference of the MAB programme in Dresden in 2011 but was clearly not implemented by AfriMAB member states.

The biosphere reserves with the biggest variety of stakeholders responsible for sustainable climate change projects are Lubombo in Eswatini, and Waterberg in South Africa. The biosphere reserves with the smallest number of responsible stakeholders were Gouritz Cluster in South Africa, whose representative stated that the biosphere reserve board member/director should be the responsible party; and Boloma-Bijagós in Guinea-Bissau, whose representative felt that it should be the responsibility of the thematic/regional network members.

6.12.3 Do you know of relevant projects implemented by other individuals or organisations in the biosphere reserves? (n=12)

Projects ran by other individuals or organisations were categorised according to country (Appendix D, Table 32, p. 277). The three countries where the most external organisations/individuals running these projects were Cameroon, DRC, and Eswatini. Malawi and Rwanda reported the smallest number of external organisations, referring to them as non-governmental organisations. The most frequently mentioned international organisations were the World Wildlife Fund (WWF), which was mentioned by three countries (25%) namely South Africa, DRC, and Mozambique, and the German Agency for International Cooperation (GIZ), which was named by Cote d'Ivoire and Burkina Faso.

6.12.4 How active are these organisations (independent from the biosphere reserve) in implementing climate change mitigation and adaptation projects in each of the different biosphere reserve zones? (n=31)

The organisations that operated independent of the biosphere reserves were considered to be "moderately" active in the core (52%), buffer (81%), and transition (45%) zones of each biosphere reserve (Appendix D, Table 33, p. 278). This means that these organisations mostly focussed their activities on the biosphere reserve's buffer zones. The buffer zone offers the ideal location where the ecosystem services that are provided by the core conservation area can be extended further into the landscape, thereby either improving its resilience to the impact of climate change, or to attempt to stem the anthropocentric challenges that could spill over into the transition zone. This is the reason OECMs, that are typically located in the buffer zone, are considered to be ideal solutions to reach the 30X30 goals.

6.12.5 What do you think the role of these organisations would be in the future (n=24)?

All respondents agreed their role would either remain the same or expand. A respondent stated that they "could have a central role in mobilising and even facilitating the awakening of the parties to respect the commitments to adapt and fight climate change", while another said they could "amplify the benefits for the vulnerable populations of the biosphere reserve and may also initiate actions similar to those developed in a biosphere reserve in other areas". Their importance was recognised by a respondent who noted that "some of them have money, expertise, and motivation. The biospheres should work with them". A respondent from Eswatini provided an example of the importance of these organisations: "in my country, there are a number of non-governmental organizations that have been working in the area of the biosphere reserve even before its nomination. These organizations have built good relations and expertise with the citizens. Thus, a synergy with the biosphere reserve would be beneficial and would fast-track its initiatives". Considering these comments, the independent organisations seem to have resources and relationships with local communities that the biosphere reserves themselves sometimes lack, something which Siders (2019) regard as essential for effective management.

6.12.6 Which stakeholder(s) must initiate and/or conduct climate change research for the results to be actively implemented? (n=33)

The top three groups deemed responsible for research in the biosphere reserve that could be implemented, were researchers (86%), consultation/advisory groups (86%), and government organisations (67%). The groups mentioned the least were thematic/regional network members (9%); and the thematic/regional network administration (18%) (Appendix D, Table 31, p. 275). Based on these results, researchers and consultants seem to contribute to organisational learning within the biosphere reserve, where present. Overall, the thematic network contributes the least to both climate change project sustainability and research, which is an area that can be improved, as each thematic network biome has a special role in climate change adaptation.

The biosphere reserve representatives who identified the largest number of stakeholder groups that should be responsible for climate change research for it to be implemented were from Boucle du Baoulé in Mali, and Yangambi in the DRC. The biosphere reserves that suggested the smallest number of stakeholder groups that should initiate research was Gouritz Cluster, which stated that it is the responsibility of the biosphere reserve board member/director group; Kogelberg, which felt the conservation area management should initiate research; and Magaliesberg, which answered that researchers should initiate research for it to be implemented. Were you aware of the details of any of the climate change scenarios mentioned above before doing this questionnaire? (n=53)

Most of the respondents were only partly aware of the potential impact of climate change (Appendix D, Table 34, p. 279). This may indicate a lack of communication between government authorities whose representatives attend crucial international events, and biosphere reserve stakeholders who must execute the initiatives decided upon at these definitive meetings.

6.13 Actor: performance measures

Four factors were used to determine the performance of actors within the biosphere reserve, namely the resilience of local livelihoods that depend upon the environment to the effects of climate change; the strength of the relationships between the biosphere reserves and local populations, businesses and organisations; the degree to which local populations, businesses and organisations are aware of the impact of that a temperature rise; and whether local companies are aware of the impact of temperature increase (n=32).

Most of the biosphere reserve representatives reported a “medium” livelihood resilience to climate change (41%); a “strong-medium” relationship between bio-

Figure 6.17 Actors' performance on the adaptive cycle.

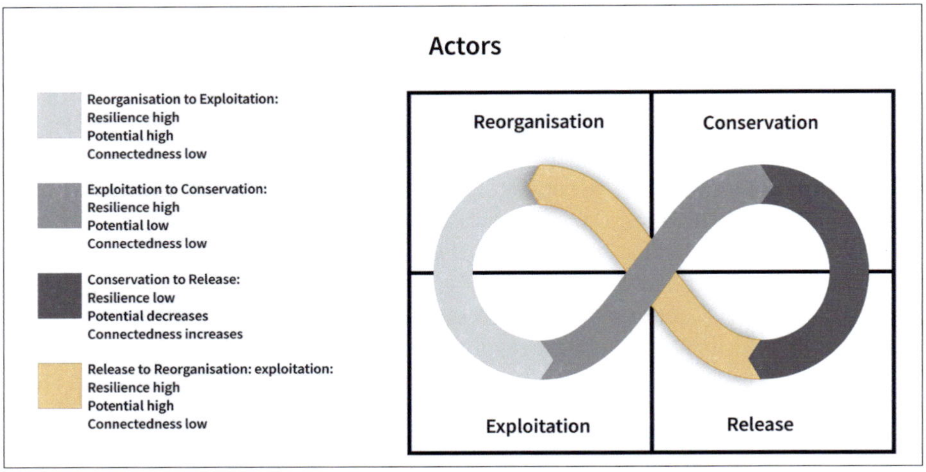

Source: Own compilation

sphere reserve, local populations, businesses and organisations (38%); they "agree" that local populations, businesses and organisations are aware of the impact of climate change (41%); and they are "neutral" about company awareness of the impact of temperature increase on them (50%) (Appendix D, Table 35, p. 280). In terms of the score of the AfriMAB actor performance, the average was 65%. The biosphere reserves that performed the best were Omo in Nigeria (100%), and Yangambi in the DRC (95%), while the lowest-scoring biosphere reserves with 40% each were Benoué, Doumba-Rey, and Waza, all in Cameroon.

Referring to the adaptive cycle, this places actor performance in the release to reorganisation phase (Figure 6.17). This means that there is generally only limited resources available for actors, but many opportunities, and that foreign influences will now have the greatest impact on the system's future (Holling & Gunderson, 2002).

Although, most of the respondents rated the Actor indicators as fairly mediocre, the low response rate failed to amount to a consensus about where the actors' performance should be placed on the adaptive cycle. From this it can be deduced that AfriMAB can improve their performance by including more stakeholders in their climate change mitigation and adaptation strategies, and that they should distribute information about climate change to the community more effectively. This supports the work by Siders (2019) on the elements needed to manage any project successfully. However, since the respondents are not fully informed about all the details of these challenges, it is crucial that the top-down communication flow should first be improved.

6.14 Interactions and outcomes

Interaction not only takes place between the elements that comprise each of the systems and subsystems mentioned above, but also between those systems and subsystems in their entirety forming the institutional structures that are needed to reach the 30X30 targets. This section highlights the consequences of those interactions, and what can be done to improve them, by looking at the cumulative performance measure scores of each section; the rated effort biosphere reserves put into limiting the global temperature increase; and the climate change scenario's that respondents believe fit their biosphere reserve's efforts. Respondents were subsequently asked what factors determine whether they pursue climate change adaptation and mitigation projects, and how AfriMAB can help limit the global temperature increase.

6.14.1 The cumulative performance measures of each section. (n=32)

The total scores of the Resource Systems, Resource Units, Governance Systems and Actors were added together and provided a total institutional performance score for each biosphere reserve. The average institution score for the entire AfriMAB network was 60% (Appendix D, Table 36, p. 283). Western and southern Africa had the best performing biosphere reserves, and eastern and central Africa had mixed results (Figure 6.18).

The overall best performing biosphere reserves were Omo in Nigeria (96%), Boucle du Baoulé in Mali (84%), and Gouritz Cluster in South Africa (76%). The worst performing biosphere reserves were Gishwati-Mukura Landscape in Rwanda (43%), Kafa in Ethiopia (52%), and Kogelberg in South Africa (52%) (Figure 6.18). When comparing the various relevant elements in these biosphere reserves, patterns emerged that may give an indication of what factors possibly contribute to their performance (Appendix D, Table 37, p. 285). Most of the better performing biosphere reserves are located in either western or southern Africa, while the worst performing biosphere reserves are either in eastern or southern Africa. The large disparity between the performance of the two South African biosphere reserves stems from the fact that South African biosphere reserves are managed by individual non-government organisations with varying levels of efficiency, and not by a government organisation or conservation authority (Stoll-Kleemann & O`Riordan, 2017). The Gouritz Cluster's greater success can in part be attributed to the USD 854,586.93 that it received from the European Union shortly after its establishment for their Jobs for Carbon Project (Gouritz Cluster, 2014). Assets such as this is one of the key factors of adaptive capacity and clearly contributed positively to the success of this biosphere reserve (Jones et al., 2011). Two of the three top performing biosphere reserves are from lower-middle income countries, while two of the three low performing biosphere reserves are from low-income countries. This may suggest that the more economically developed a country is, the more likely it is that the biosphere reserves will perform better, as there are generally more tangible and intangible assets available to support conservation efforts.

Apart from the Gouritz Cluster, the older biosphere reserves tend to perform better than the less established ones. The Gishwati-Mukura Landscape in Rwanda (est. 2020) is a relatively new biosphere reserve in the AfriMAB network, and while it is not performing well now, its performance will more than likely improve over time as it already has two active climate change projects. The successful biospheres reserves have experience, knowledge and institutional memory which the younger biosphere reserves do not have, which is key to their achievements (Brooks et al., 2004). On the other hand, Cuong et al. (2017a) found that younger biosphere reserves tend to be more successful, because of the post-Seville model they follow. This might mean that the older biosphere reserves mentioned here, have upgraded their governance structure to better follow the post-Seville model, making them successful, but a deeper analysis is needed to determine if this is in fact the case.

Results indicate that the more successful biosphere reserves run at least one climate change project every three years, while the worst performing biosphere reserves generally had either single projects lasting under a year, or none at all. The longer projects would potentially have a greater impact on the landscape, and moreover, is indicative of a stable biosphere reserve management structure that is able to implement and maintain long-term projects.

The better performing biosphere reserves all included some research projects on climate change impacts or a CO_2 assessment, but the worst performing biosphere reserves did not. Cinner et al. (2018) rightfully said that learning is a key element in adaptive capacity of an organisations. Based on these biosphere reserves' success they apparently put the answers obtained through the research to good use. Projects that are well funded also seem to contribute to the biosphere reserves performance, as two of the most successful ones reported that they received funding of between USD 10 per km^2 and USD 19 per km^2. This once again proves that good performance is enhanced by adequate assets to execute the projects.

Figure 6.18 Biosphere reserve performance by category.

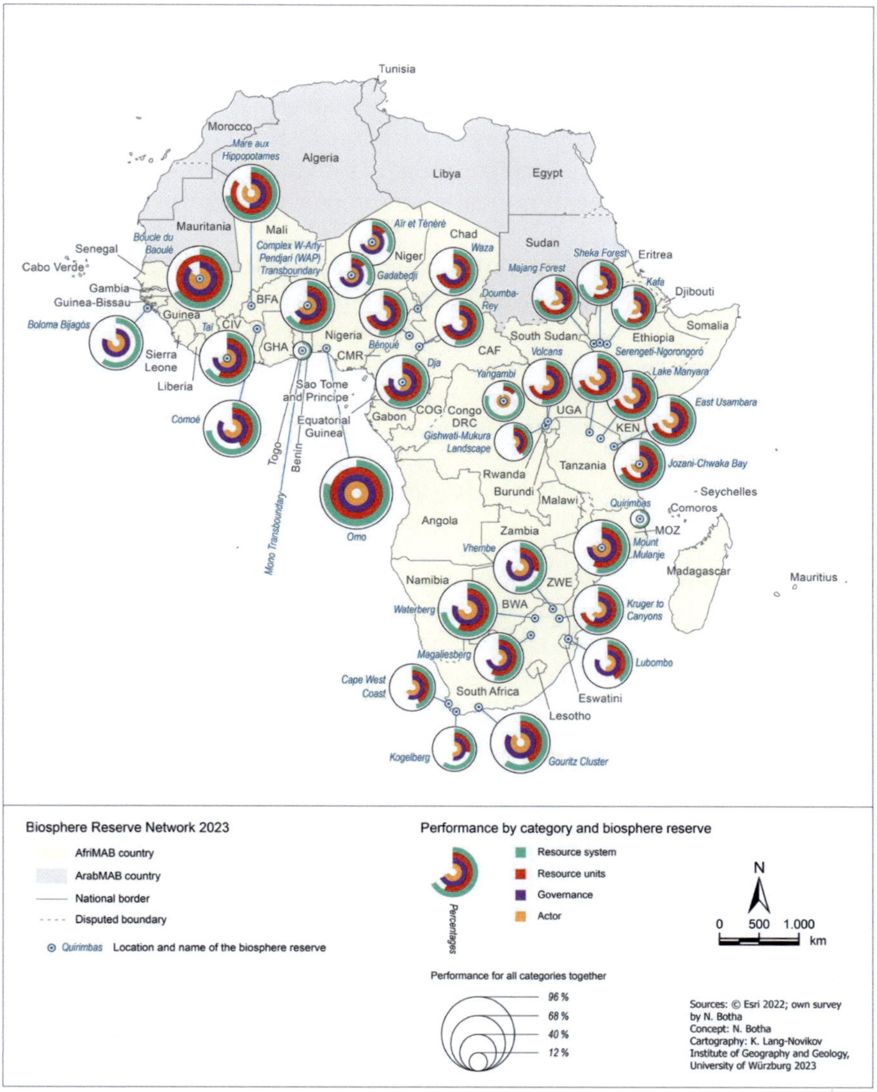

Source: Own compilation

Interestingly, an aspect that did not seem to contribute to success, was the institutionalisation scores. Two of the three best performing biosphere reserves did not have a budget for climate change projects, but this is probably because their funding came from external sources. These same biosphere reserves reported having either a leading, partner, or coordinating role in the climate change projects, so it is possible that the available funds are managed by their external partners. The struggling biosphere reserves each had only one role in the projects, but never as the leader. This might indicate a lack of initiative on their part, which is also an important element that is needed when promoting adaptive capacity (Cinner et al. (2018).

Most funding comes from international and local sources, which agrees with findings by (Cuong et al., 2017b), which also showed a lack of investment by national government authorities in biosphere reserve projects. Biosphere reserves that tended to perform well rather had organisations that were actively and financially involved in the core and buffer zones of the biosphere reserve, but especially focused on the transition zone. This makes sense, as these are the areas where people be the hardest hit by climate change; this is where they live, and this is generally where most of the landscape degradation and greenhouse gas emissions occur.

6.14.2 In your opinion, is the organisation you represent in the UNESCO Biosphere Reserve doing as much as possible towards limiting global temperature increase to below 2°C? (n=30)

In answer to this question, most respondents ranked their biosphere reserve's efforts at seven (27%) or six (17%), the average rank being six (Appendix D, Table 38, p. 287) (Figure 6.19). This correlates with the average performance of biosphere reserves.

The biosphere reserves whose efforts to limit the increase in global temperatures were the highest, are Sahamalaza-Iles Radama and Tsimanampesotse-Nosy Ve Androka in Madagascar. Unfortunately, the representative(s) from these biosphere reserves did not complete the second questionnaire that contained the performance measures. The lowest ranking biosphere reserve was Cape Winelands in South Africa, and Jozani-Chwaka Bay in Tanzania. Cape Wineland's total performance scored 55%, and Jozani-Chwaka Bay scored 63%, which means that while they may consider their effort to be insufficient, their total performance is average.

Figure 6.19 Ranking of efforts to limiting global temperatures to below 2°C from least to most effort.

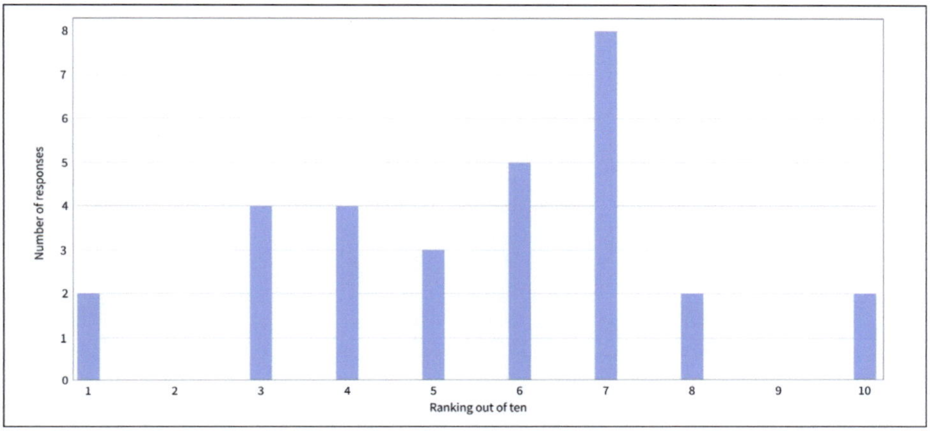

Source: Own compilation

Omo (Nigeria) ranked their effort a seven, which is realistic considering their overall performance, Boucle du Baoulé ranked their effort a four and Gouritz Cluster a five, which implies that although they perform well, they feel they can do more. Gishwati-Mukura Landscape ranked their efforts a seven, but this is incongruent with their performance as a biosphere reserve. This may be because their work has not yet translated into results because of their age. However, an in-depth analysis is needed to determine the truth of this assumption.

6.14.3 In which climate change scenario does your organisation's efforts fit?

Most respondents (67%) stated that their organisation's efforts fit into Scenario 2, limiting global temperature increase to below 2°C (Appendix D, Table 39, p. 288) (n=56). This question was repeated in the second questionnaire (n=25). The results indicate a major perspective change with an improvement of organisations' efforts (Figure 6.20). Most respondents (60%) indicated in the second questionnaire that their organisations' efforts fit into Scenario 3, where average global temperature would remain below 1.5°C. This shows that the consensus on this matter had shifted and that respondents become more optimistic over the six-month period (Appendix D, Table 40, p. 290). There may be many reasons for this,

Figure 6.20 Comparison of organisations' effort in terms of the climate change scenarios between the first and second questionnaires.

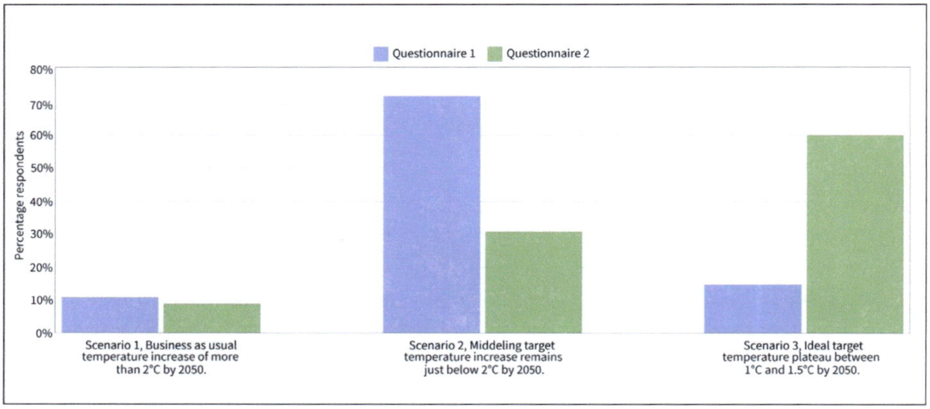

Source: Own compilation

for example, the second questionnaire did not contain facts about each scenario again, which may have been sobering in the first questionnaire. Also, the 30X30 goal might have contributed to a more positive outlook. Furthermore, biosphere reserves began to implement more climate change related projects in the six months between the two questionnaires. These explanations are however purely speculative, and more information is needed to provide a conclusive reason for this shift.

Respondents that now believed that their organisations were not doing enough to curb the global temperature increase were from WAP in Benin, Burkina Faso and Niger, and Omo in Nigeria. This is interesting, considering that Omo's overall institutional performance was 96% with an organisational effort ranking of seven. However, if the potential exists for them to do more, their biosphere reserve will only improve if they take actions towards it.

6.14.4 What are the factors that determine whether a climate change project will be pursued by their organisation? (n=29)

Respondents most frequently mentioned funding (59%), stakeholder involvement (38%), and capacity building of the biosphere reserve management structure (17%) as the factors that will determine whether a project will be pursued (Appendix D, Table 41, p. 291). These results support the findings by Cinner et al. (2018) on the key elements needed to build adaptive capacity. The four biosphere reserves that provided the most factors were WAP in Benin, Burkina Faso, and Niger; Taï and Comoé in Côte d'Ivoire; and Boucle du Baoulé in Mali, which are, interestingly, all located in western Africa.

6.14.5 Do you have any further suggestions on how the UNESCO Biosphere Reserve network can help limit the global temperature increase to below 1.5°C? (n=41)

Respondents' answers to this question in the first questionnaire were organised according to the categories of the social-ecological systems framework, and presented as such in the second questionnaire (Appendix D, Table 42, p. 295). The respondents were asked to indicate which of the elements mentioned in the first questionnaire were the most important within each provided category (n=23). Overall, each category received an average of 13 votes from biosphere representatives, which is more than 65% of the votes. This can be accepted as a majority consensus.

In the Governance System category, the most indicated (74%) element was capacity building of biosphere reserves to maintain their role as learning centres for sustainable development, by reconciling conservation and development, which means that AfriMAB representatives recognise that they have problems with their logistic support function. To address this shortcoming, the MAB programme should consider offering an accredited course to biosphere reserve managers.

In the Resource System category, the most mentioned (83%) element was "improving conservation of the buffer zone and transition of the reserve", which is needed to reach the 30X30 target, and to specifically improve environmental resilience to climate change. The fact that respondents indicated this, means there are more opportunities available to improve conservation in these zones.

Under Resource Units, the element that was highlighted the most (87%) was: "become more involved with the acquisition of funding for biosphere reserves that are reaching their goals". This study confirmed that funding is an important issue not only for individual projects, but that under the right conditions it could kickstart the performance of the entire biosphere reserve (e.g., Gouritz Cluster).

In the Actor category, two major elements (86% each), were pointed out, namely: "raise awareness of biosphere reserves among the public and decision-makers" and "increase the participation of local communities living near biosphere reserves in the management and conservation of natural resources and encourage them to reduce their CO_2 emissions and promote sustainable agriculture". This study also found that there is a need to raise awareness within especially national governments of the possible role of biosphere reserves. The involvement of more local stakeholders is exactly what is needed for biosphere reserves to reach the 30X30 goals.

In terms of Interactions, “evaluation of the amount of CO_2 sequestered by tropical forest biosphere reserves to proceed with financial compensation. This financial windfall could be used for activities compatible with the conservation of natural resources for the benefit of locals”, was mentioned the most (70%). This was already suggested by MAB (2011), but it was not widely implemented in the AfriMAB network.

The element within the Outcome category that received the most votes (65%) was to “develop a green economy in biosphere reserves by creating alternative livelihood projects”. While some authors within the German National Unesco Commission (2015) believed that biosphere reserves should move away from green economy projects, respondents indicated that the more economically developed a country or biosphere reserve is, the more resources are available to enhance conservation measures. However, developing green economies will be less resource intense than traditional forms of development, and will promote both the natural conservation and the sustainable development functions of the biosphere reserves.

6.15 Concluding remarks

The results of the two online questionnaires were provided in this chapter and organised into the social-ecological systems framework categories, namely the Social, Economic, and Political Setting, Related Ecosystems, Resource Systems, Resource Units, Governance Systems, Actors, Interactions, and Outcomes. The results of each of these sections, apart from Interactions and Outcomes, were then discussed in terms of either future predictions of the category, or its performance.

7 Conclusion

In exploring climate change mitigation and adaptation strategies in AfriMAB, this study aimed to investigate how this network could position itself in its current institutional structure to better contribute to the CBD's 30X30 target and specifically improve their environment's resilience to climate change.

To accomplish this, several objectives were formulated and addressed:

1. *Broaden the theoretical understanding of how international climate change, nature conservation and MAB events have influenced biosphere reserves' creation, evolution, and current institutional structures, by reviewing relevant literature.*
   It was found that climate change concerns already surfaced during the Industrial Revolution (Weart, 2008), which sparked a global debate about the impact of human activity on the atmosphere and the global trend of increased temperatures (May, 2020). Industrialisation also led to mechanized agriculture and consequently urban migration, transforming natural landscapes into areas which prioritized industry over biodiversity (Antrop, 2004). This brought about landscape degradation, and a loss of biodiversity and ecosystem services, which created a need for environmental conservation movements and protected areas such as national parks. Over time, protected areas became isolated in man-made landscapes (Bernhard Gissibl et al., 2012), leading to international concerns and the establishment of biosphere reserves to try to find a balance between development and conservation (McCormick, 1995). Evolving with international targets, biosphere reserves formed a complex system with diverse management structures.
2. *Determine which is the best theoretical framework to analyse biosphere reserves' complex institutional structures through a thorough literature review on systems thinking and social-ecological systems.*
   It was established that the most appropriate theoretical framework for this study was an adapted version of the Ostrom (2009) social-ecological systems framework which included the Delphi research methods (Vogel et al., 2019), and performance-measuring elements that relate to the adaptive cycle (Castell & Schrenk, 2020).
3. *Establish the elements within the chosen social-ecological system framework that will be used as measures of each biosphere reserve's institutional structures, which are based on principal guiding biosphere reserve documents.*

These documents included the: "Statutory framework of the World Network of Biosphere Reserves"; "Lima Action Plan for UNESCO's Man and the Biosphere (MAB) Programme and its World Network of Biosphere Reserves (2016-2025)"; "The management manual for UNESCO Biosphere Reserves in Africa"; and "For life, for the future: Biosphere reserves and climate change". Relevant measuring elements relating to the institutional structures of biosphere reserves and their efforts to address climate change were extracted from these documents according to the frameworks' seven categories, namely the social, economic, and political settings; related ecosystems; resource systems; resource units; governance systems; and actors.

4. *Describe the methodology and methods used during this study that ensured that it was reliable, valid, and replicable and able to contribute to the academic body of knowledge on climate change actions, and biosphere reserves.*
   This study was an empirical mixed-methods research endeavour utilizing a cross-sectional research design. The data collection involved deploying two online questionnaires to the AfriMAB survey population, spaced six months apart. Before distribution, the questionnaires underwent expert review and were used in a pilot study. The primary sampling method employed was non-probability sampling. The data analysis encompassed quantitative, qualitative, and spatial analysis techniques.
5. *Provide a brief introduction to the geographical and environmental context for the countries in sub-Sahara Africa and the African Regional Biosphere Reserve Network (AfriMAB).*
   It was determined that the AfriMAB Biosphere Reserve network is facing a formidable challenge in a region fraught with vulnerabilities to climate change. Navigating the quest for effective climate solutions within the constraints of limited resources and the ever-shifting international landscape poses a unique challenge not commonly encountered by other regions.
6. *Distribute a multi-round online questionnaire to 100 members of the AfriMAB network to determine what the current state of African biosphere reserves' climate change action is, and their performance.*
   The first survey period was from 6 April to 15 June 2022, and the second survey period six months later from 19 January to 30 March 2023. Seventy respondents completed the initial questionnaire, with 63 expressing their willingness to participate in the subsequent round. This subset, comprising the new survey population, eventually saw

active involvement from 38 respondents. In the first and second rounds 34 (37%) and 35 (38%) AfriMAB Biosphere Reserves were respectively represented, only 26 (28%) participated in both.

7. *Apply the chosen framework and make suggestions on how the AfriMAB Biosphere Reserve network can collectively position itself in its current institutional structures to better contribute to climate change adaptation and mitigation.*
   The ensuing recommendations are systematically organized within each category of the social-ecological systems framework.

Social, economic, and political settings:

First, to improve the overall participation in UN activities and reduce the countries' climate change vulnerability, AfriMAB countries must grow economically, preferably by practising green economic development. Second, the UN should lower the cost of participation in their events by auditing their various organisations, funds, and programmes to ensure that their agendas do not overlap and synchronise international events to reduce the number and frequency of meetings. This will enable more relevant representatives, especially those from less affluent regions for whom the cost of attending these events is prohibitive, to attend vitally important gatherings. Third, the UN should address the free-rider problem at their events, and reevaluate who the relevant participants are, to gain meaningful results.

Related ecosystems:

First, biosphere reserves should focus on the restoration and enhancement of key biomes that will help their society to adapt to the impact of climate change, including wetlands, forests, savannahs, and marine biomes. Second, biosphere reserves should focus their efforts on their core and transition zones to prevent anthropogenic knock-on effects from reaching the transition zones. Third, considering the importance of landscape restoration in the face of climate change, the MAB programme should contemplate holding its World Network of Biosphere Reserve Meeting, closer to the date of the CBD's international biodiversity decade target-setting conference. Fourth, the thematic networks should become more active.

Resource systems:

First, the MAB programme must re-excite older biosphere reserves about newer targets, such as climate change adaptation and mitigation. Second, biosphere reserves must conduct studies on climate change impact and CO_2 emission before projects are initiated. Without specific knowledge about their unique situation, projects may be futile, and waste already scarce funds. Third, an early influx of assets (both tangible and intangible) in young biosphere reserves will help to establish them in the landscape and positively impact their ability to implement climate change mitigation and adaptation in the future.

Resource units:

First, AfriMAB Biosphere Reserves must include more sustainable development projects relating to climate change. Second, AfriMAB must encourage and conduct more research in their biosphere reserves, especially relating to the social aspects. Third, the MAB programme must train biosphere reserves to better implement their logistic support function. Fourth, biosphere reserves must continue implementing multiprong climate change projects with a variety of partners, while adopting various roles in climate change projects. Fifth, the MAB Programme must consider holding its World Network of Biosphere Reserves meeting every five years, instead of every 10 years, to maintain momentum and enthusiasm for climate change projects. Sixth, when determining the success of climate change in the World Network of Biosphere Reserves, the MAB programme might need to include the percentage of project cover in the biosphere reserve and the duration of the project as well. Seventh, AfriMAB Biosphere Reserves must have at least one climate change project every three years to elevate it success to the level of the best-performing biosphere reserves in their network. Eight, biosphere reserves must obtain enough funding for each project; this study suggests an amount of at least USD 20 per km^2 per year.

Governance systems:

First, the MAB programme urgently needs to encourage national government recognition and investment in biosphere reserves, as the IUCN did with their protected area categories. Second, biosphere reserves must therefore budget for climate change projects. Third, these projects must be initiated locally, but partners who get involved at this stage, will have a marked positive impact on the future of the biosphere reserve.

Actors:

First, member states of the AfriMAB network must improve their communication with their biosphere reserves. Second, biosphere reserves must prioritise their communication with local communities, organisations, and businesses. Third, climate change projects must be driven by government organisations or biosphere reserve management to be sustainable, while research must be initiated by researchers, advisors, or consultants to be actively implemented.

In employing the above-mentioned suggestions, the UNESCO AfriMAB network could position itself in its current institutional structure to better contribute to the CBD's 30X30 target and specifically improve the environment's resilience to climate change. The results of this study will hopefully help managers and decision-makers to recognise possible areas of improvement and encourage them to take appropriate actions. This study addressed a knowledge gap about the current state of climate change actions in the AfriMAB network, one of the most vulnerable networks to climate change in the world.

Despite diligent efforts to ensure the comprehensiveness of this study, certain limitations persist. The online format introduced the challenge of survey fatigue among respondents, potentially leading to more superficial responses. Additionally, the diminishing response rate across questionnaire rounds restricted the depth of exploration on certain topics. The absence of repeat responses from the same biosphere reserves in multiple survey rounds negatively affected the study.

Although this study has limitations, it opens exciting avenues for future research. Both the results and the discussion sections raise questions that deserve further investigation, pointing the way for future inquiries. One such area of research is exploring how well the concept of biosphere reserves is incorporated into the laws of member states within both AfriMAB and the World Network of Biosphere Reserves. Another compelling research opportunity is applying the same methods to all regional networks of the MAB program. This would allow for a comprehensive comparison of their effectiveness in addressing climate change challenges.

This study delved into the AfriMAB Biosphere Reserve network, exploring its fascinating social and ecological dynamics across time and space. We examined its efforts to address climate change mitigation and discovered that by providing comprehensive support tailored to each reserve's unique needs, it will be possible to significantly boost their effectiveness. Such targeted reinforcement will empower these biosphere reserves to make long-term, sustainable contributions on a landscape scale, and to play a crucial role in the fight against climate change. The benefits of this will extend beyond environmental improvements but will also positively impact societal well-being and highlight the powerful link between man and nature.

Bibliography

ACAR, S., & YELDAN, E. (2019). *Handbook of green economics*: Elsevier Science.

ACKOFF, R. L. (1971). Towards a system of systems concepts. *Management Science, 17*(11), 661-671. doi: 10.1287/mnsc.17.11.661.

ACKOFF, R. L. (2010). *Systems thinking for curious managers*: Triarchy Press.

ADDANEY, M., JARBANDHAN, D. B., & DUMENU, W. K. (2023). *Climate change in Africa: Adaptation, resilience, and policy innovations*: Springer Nature Switzerland.

ADEYEMI, O. (2018). Analysis and prediction of land cover change of a biosphere reserve using geospatial tools: A case study of Omo biosphere, Nigeria. *OIDA International Journal of Sustainable Development, 10*(4), 35-48.

ADEYEMI, O., CHIRWA, P. W., & BABALOLA, F. D. (2022). Assessing local people's perceptions and preference for ecosystem services to support management plan in Omo Biosphere Reserve, Nigeria. *Environmental Development, 43*, 100738.

ADGER, W. N., HUGHES, T. P., FOLKE, C., CARPENTER, S. R., & ROCKSTRÖM, J. (2005). Social-ecological resilience to coastal disasters. *Science, 309*(5737), 1036-1039. doi: 10.1126/science.1112122.

ADLER, M., & ZIGLIO, E. (1996). *Gazing into the oracle: The Delphi method and its application to social policy and public health*: Jessica Kingsley Publishers.

AFRIMAB. (2010). *African biosphere reserve network: AfriMAB charter*. Nairobi.

AFRICAN WILDIFE FOUNDATION, A. W. (2016). *2016 Annual Report*. Retrieved from https://www.awf.org/about/resources/awf-2016-annual-report.

AGGARWAL, P., VYAS, S., THORNTON, P., & CAMPBELL, B. M. (2019). How much does climate change add to the challenge of feeding the planet this century? *Environmental Research Letters, 14*(4), 043001-043001. doi: 10.1088/1748-9326/AAFA3E.

ALJAZEERA. (2023, 8 Dec 2023). OPEC rallies members against fossil fuels phase out at COP 28, News. *Aljazeera*. Retrieved from https://www.aljazeera.com/news/2023/12/8/opec-rallies-members-against-fossil-fuels-phase-out-at-cop-28.

ALLAN, J., BANSARD, J., AKANLE, T., ENI-IBUKUN, LUOMI, M., SPENCE, C., & TAN, J. M. (2022). Summary of the Sharm El-Sheikh climate change conference: 6-20 November 2022. *Earth Negotiations Bulletin*. Retrieved from https://enb.iisd.org/sharm-el-sheikh-climate-change-conference-cop27-summary.

ALTHOR, G., WATSON, J. E. M., & FULLER, R. A. (2016). Global mismatch between greenhouse gas emissions and the burden of climate change. *Scientific Reports, 6*(1), 20281. doi: 10.1038/srep20281.

AMARUZAMAN, S., TRONG HOAN, D., CATACUTAN, D., LEIMONA, B., & MALESU, M. (2022). Polycentric environmental governance to achieving sdg 16: Evidence from Southeast Asia and Eastern Africa. *Forests, 13*(1). doi: 10.3390/f13010068.

AMER, W., ASHONG, S., & TIOMOKO, D. (2015). *Management manual for UNESCO biosphere reserves in Africa: Field version*: German Commission for UNESCO.

ANTROP, M. (2004). Landscape change and the urbanization process in Europe. *Landscape and urban planning, 67*(1-4), 9-26. doi: 10.1016/S0169-2046(03)00026-4.

APPS, J., & KASSULKE, N. (2023). *Planting an idea: Critical and creative thinking about environmental issues*: Fulcrum Publishing.

ARIHANT. (2021). *OPSC general studies paper 1 (for Odisha civil service preliminary exams) 2021*: Arihant Publications India limited.

ARRHENIUS, S., & HOLDEN, E. S. (1897). On the influence of carbonic acid in the air uppon the temperature of the Earth. *Publications of the Astronomical Society of the Pacific, 9*(54), 14-24.

AUBER, A., WALDOCK, C., MAIRE, A., GOBERVILLE, E., ALBOUY, C., ALGAR, A. C., MCLEAN, M., BRIND'AMOUR, A., GREEN, A. L., TUPPER, M., VIGLIOLA, L., KASCHNER, K., KESNER-REYES, K., BEGER, M., TJIPUTRA, J., TOUSSAINT, A., VIOLLE, C., MOUQUET, N., THUILLER, W., & MOUILLOT, D. (2022). A functional vulnerability framework for biodiversity conservation. *Nat Commun, 13*(1), 4774. doi: 10.1038/s41467-022-32331-y.

AYUGI, B., ERESANYA, E. O., ONYANGO, A. O., OGOU, F. K., OKORO, E. C., OKOYE, C. O., ANORUO, C. M., DIKE, V. N., ASHIRU, O. R., DARAMOLA, M. T., MUMO, R., & ONGOMA, V. (2022). Review of meteorological drought in Arica: Historical trends, impacts, mitigation measures, and prospects. *Pure and Applied Geophysics, 179*(4), 1365-1386. doi: 10.1007/s00024-022-02988-z.

BALNAVES, M., & CAPUTI, P. (2001). *Introduction to quantitative research methods: An investigative approach*: SAGE Publications.

BARRACLOUGH, A. D., & MÅREN, I. E. (2022). *The role of UNESCO biosphere reserves in the implementation of the convention on biological diversity's post-2020 global biodiversity framework: Policy brief*: UNESCO.

BATIE, S., & MERCURO, N. (2008). *Alternative institutional structures: Evolution and impact*: Taylor & Francis.

BATISSE, M. (1982). The biosphere reserve: A tool for environmental conservation and management. *Environmental Conservation, 9*(2), 101-111. doi: 10.1017/S0376892900019937.

BATISSE, M. (1985). Action plan for biosphere reserves. *Environmental Conservation, 12*(1), 17-27. doi: 10.1017/S0376892900015113.

Baxter, K., Courage, C., & Caine, K. (2015). Preparing for your user research activity. In K. Baxter, C. Courage, & K. Caine (Eds.), *Understanding your users (second edition)* (114-156). Boston: Morgan Kaufmann.

Beasley, R. (2012). 4.3.1 The barriers to systems thinking. *INCOSE International Symposium, 22*, 517-531. doi: 10.1002/j.2334-5837.2012.tb01353.x.

Becken, S., & Job, H. (2014). Protected Areas in an era of global-local change. *Journal of Sustainable Tourism, 22* (4), 507-527. doi: 10.1080/09669582.2013.877913.

Beer, S. (1979). *The heart of enterprise*: Wiley.

Behl, D. V., & Ferreira, S. (2014). Systems thinking: An analysis of key factors and relationships. *Procedia Computer Science, 36*, 104-109. doi: 10.1016/j.procs.2014.09.045.

Beiderbeck, D., Frevel, N., von der Gracht, H. A., Schmidt, S. L., & Schweitzer, V. M. (2021). Preparing, conducting, and analyzing Delphi surveys: Cross-disciplinary practices, new directions, and advancements. *MethodsX, 8*, 101401. doi: 10.1016/j.mex.2021.101401.

Bell, S. (2009). Experimental design. In R. Kitchin & N. Thrift (Eds.), *International Encyclopedia of Human Geography* (pp. 672-675). Oxford: Elsevier.

Bentahar, Z. (2011). Continental drift: The disjunction of north and sub-Saharan Africa. *Research in African Literatures, 42*(1), 1-13. doi: 10.2979/reseafrilite.2011.42.1.1.

Berhanu, K., & Teshome, E. (2018). Opportunities and challenges for wildlife conservation: The case of Alatish National Park, Northwest Ethiopia. *African Journal of Hospitality, Tourism and Leisure, 7*(1), 1-13.

Berkes, F., Folke, C., & Colding, J. (1998). *Linking social and ecological systems: Management practices and social mechanisms for building resilience*: Cambridge University Press.

Bertalanffy, L. V. (1950). An outline of general system theory. *The British Journal for the Philosophy of Science, 1*(2), 134-165. doi: 10.1093/bjps/I.2.134.

Biggs, R., Gordon, L., Raudsepp-Hearne, C., Schlüter, M., & Walker, B. (2015a). Principle 3: Manage slow variables and feedbacks. In: Biggs, R. (Oonsie), M. Schlüter, & M.L. Schoon (Eds.), *Principles for Building Resilience: Sustaining Ecosystem Services in Social-Ecological Systems* (pp. 105-141). Cambridge: Cambridge University Press. doi: 10.1017/CBO9781316014240.006.

Binder, C. R., Hinkel, J., Bots, P. W. G., & Pahl-Wostl, C. (2013). Comparison of frameworks for analyzing social-ecological systems. *Ecology and Society, 18*(4). doi: 10.5751/ES-05551-180426.

Bires, Z., & Raj, S. (2019). Determinants of environmental conservation in Lake Tana Biosphere Reserve, Ethiopia. *Heliyon, 5*(7). doi: 10.1016/j.heliyon.2019.e01997.

BITETTI, M., BITETTI, T., & CALVERT, C. K. (2013). *The aviation history of Greater Riverside*: Arcadia.

BLAHNA, D. J., ASAH, S. T., DEAL, R. L. (2017). An ecosystem services framework. In D. H. Olson, & B. Van Horne (Eds), *People, forests, and change*. Washington: Island Press.

BLAIKIE, N., & PRIEST, J. (2017). *Social research: Paradigms in action*: Wiley.

BLAIKIE, P. (2006). Is small really beautiful? Community-based natural resource management in Malawi and Botswana. *World Development, 34*(11), 1942-1957. doi: 10.1016/j.worlddev.2005.11.023.

BMZ. (2022). Good governance. Retrieved from https://www.bmz.de/de/entwicklungspolitik/good-governance.

BOTHA, E. (2023). Stakeholder perceptions of sustainability and possible behaviour in a biosphere reserve. *Sustainable Development, 31*(5), 3843-3856. doi: 10.1002/sd.2629.

BOTHA, N., JOB, H., & KIMARIO, F. (2021). Potential and challenges of the Serengeti-Ngorongoro Biosphere Reserve, Tanzania. *Eco.mont (Journal on Protected Mountain Areas Research), 13*, 27-37. doi: 10.1553/eco.mont-13-sis27.

BRAUCH, H. G., SPRING, Ú. O., MESJASZ, C., GRIN, J., KAMERI-MBOTE, P., CHOUROU, B., DUNAY, P., & BIRKMANN, J. (2011). *Coping with global environmental change, disasters and security: Threats, challenges, vulnerabilities and risks*: Springer Berlin.

BRAUN, V., & CLARKE, V. (2021). *Thematic analysis: A practical guide*: SAGE Publications.

BRENNER, L., & JOB, H. (2012). Challenges to actor-oriented environmental governance: Examples from three Mexican biosphere reserves. *Tijdschrift voor Economische en Sociale Geografie, 103*(1), 1-19. doi: 10.1111/j.1467-9663.2011.00671.x.

BRENNER, L., & JOB, H. (2022). Reviewing the participatory management of UNESCO biosphere reserves: What do we miss by ignoring local academic knowledge in Mexico? *Ambio, 51*(7), 1726-1738. doi: 10.1007/s13280-021-01672-1.

BRÉTHAUT, C., GALLAGHER, L., DALTON, J., & ALLOUCHE, J. (2019). Power dynamics and integration in the water-energy-food nexus: Learning lessons for transdisciplinary research in Cambodia. *Environmental Science & Policy, 94*, 153-162. doi: 10.1016/j.envsci.2019.01.010.

BRIDGEWATER, P. (2016). The man and biosphere programme of UNESCO: Rambunctious child of the sixties, but was the promise fulfilled? *Current Opinion in Environmental Sustainability, 19*, 1-6. doi: 10.1016/j.cosust.2015.08.009.

BROECKER, W. S. (1975). Climatic change: Are we on the brink of a pronounced global warming? *Science, 189*(4201), 460-463. doi: 10.1126/science.189.4201.460.

BRYMAN, A. (2016). *Social research methods*: Oxford University Press.

BUJANG, M. A., OMAR, E. D., & BAHARUM, N. A. (2018). A review on sample size determination for Cronbach's Alpha test: A simple guide for researchers. *Malays J Med Sci, 25*(6), 85-99. doi: 10.21315/mjms2018.25.6.9.

BURNS, M. (2008). *Exploring sustainability science: A Southern African perspective*: African Sun Media.

CACCIATTOLO, M. (2015). Ethical considerations in research. In M. Vicars, S. Steinberg, T. McKenna, & M. Cacciattolo (Eds), The praxis of English language teaching and learning. Rotterdam: SensePublishers. doi: 10.1007/978-94-6300-112-0_4.

CAPALDI, N., IDOWU, S. O., SCHMIDPETER, R., & BRUECKNER, M. (2019). *Responsible business in uncertain times and for a sustainable future*: Springer International Publishing.

CARFORA, A., & SCANDURRA, G. (2019). The impact of climate funds on economic growth and their role in substituting fossil energy sources. *Energy Policy, 129*, 182-192. doi: 10.1016/j.enpol.2019.02.023.

CARIUS, F., & JOB, H., (2019). Community involvement and tourism revenue sharing as contributing factors to the UN Sustainable Development Goals in Jozani–Chwaka Bay National Park and Biosphere Reserve, Zanzibar. *Journal of Sustainable Tourism, 27* (6), 826-846. doi: 10.1080/09669582.2018.1560457.

CARLSON, M. D., & MORRISON, R. S. (2009). Study design, precision, and validity in observational studies. *J Palliat Med, 12*(1), 77-82. doi: 10.1089/jpm.2008.9690.

CARPENTER, S., & BROCK, W. (2008). Adaptive capacity and traps. *Ecology and Society, 13*. doi: 10.5751/ES-02716-130240.

CARPENTER, S., & COTTINGHAM, K. (1997). Resilience and restoration of lakes. *Conservation ecology, 1*(2).

CARSON, R. (1962). *Silent spring*: Routledge.

CASSIA, R., NOCIONI, M., CORREA, N., & LAMATTINA, L. (2018). Climate change and the impact of greenhouse gasses: CO_2 and NO, friends and foes of plant oxidative stress. *Frontiers in Plant Science, 9*. doi: 10.3389/fpls.2018.00273.

CASTELL, W. Z., & SCHRENK, H. (2020). Computing the adaptive cycle. *Scientific Reports, 10*(1), 18175. doi: 10.1038/s41598-020-74888-y.

CASTILLO, A., BULLEN-AGUIAR, A. A., PEÑA-MONDRAGÓN, J. L., & GUTIÉRREZ-SERRANO, N. G. (2020). The social component of social-ecological research: Moving from the periphery to the center. *Ecology and Society, 25*(1). doi: 10.5751/ES-11345-250106.

CBD. (2011). *Strategic plan for biodiversity 2011–2020 and the Aichi targets*: CBD.

CBD. (2018). *Protected areas and other effective area-based conservaton measures*: CBD.

CBD. (2022a). By 2030: Protect 30% of earth's lands, oceans, coastal areas, inland waters; reduce by $500 billion annual harmful government subsidies; cut food waste in half [Press release]. Retrieved from https://www.cbd.int/article/cop15-cbd-press-release-final-19dec2022

CBD. (2022b). COP15: Nations adopt four goals, 23 targets for 2030 in landmark UN biodiversity agreement [Press release]. Retrieved from https://www.cbd.int/article/cop15-cbd-press-release-final-19dec2022.

CHANDRAKAR, A. K., YADAV, K., KUMAR, V., & GUPTA, N. (2016). *Scenario of biodiversity conservation in India: An overview*: Phonton.

CHERIAN, A. (2015). *Energy and global climate change: Bridging the sustainable development divide*: Wiley.

CHERKASSKII, B. L. (1988). The system of the epidemic process. *Journal of hygiene, epidemiology, microbiology, and immunology, 32*(3), 321-328.

CHOWDHURY, R. (2019). *Systems thinking for management consultants: Introducing holistic flexibility*: Springer Singapore.

CHOWDHURY, R. (2023). Methodological flexibility in systems thinking: Musings from the standpoint of a systems consultant. *Systemic Practice and Action Research, 36*(1), 59-86. doi: 10.1007/s11213-022-09597-w.

CIA. (2023). The world factbook: Cote d'ivoire. Retrieved from https://www.cia.gov/the-world-factbook/countries/cote-divoire/#energy.

CINNER, J. E., ADGER, W. N., ALLISON, E. H., BARNES, M. L., BROWN, K., COHEN, P. J., GELCICH, S., HICKS, C. C., HUGHES, T. P., LAU, J., MARSHALL, N. A., & MORRISON, T. H. (2018). Building adaptive capacity to climate change in tropical coastal communities. *Nature Climate Change, 8*(2), 117-123. doi: 10.1038/S41558-017-0065-X.

CINNER, J. E., & BARNES, M. L. (2019). Social dimensions of resilience in social-ecological systems. *One Earth, 1*(1), 51-56. doi: 10.1016/j.oneear.2019.08.003.

CLAPP, B. W. (2014). *An environmental history of Britain since the industrial revolution*: Taylor & Francis.

CLARK, A. M. (1998). The qualitative-quantitative debate: Moving from positivism and confrontation to post-positivism and reconciliation. *Journal of advanced nursing, 27*(6), 1242-1249. doi: 10.1046/j.1365-2648.1998.00651.x.

CLARKE, C. (2006). *Automotive production systems and standardisation: From Ford to the case of Mercedes-Benz*: Physica-Verlag HD.

COETZER, K. L., WITKOWSKI, E. T. F., & ERASMUS, B. F. N. (2014). Reviewing biosphere reserves globally: Effective conservation action or bureaucratic label? *Biological Reviews, 89*(1), 82-104. doi: 10.1111/brv.12044.

COLDING, J., & BARTHEL, S. (2019). Exploring the social-ecological systems discourse 20 years later. *Ecology and Society, 24*(1). doi: 10.5751/ES-10598-240102.

CROSSAN, F. (2003). Research philosophy: Towards an understanding. *Nurse Researcher, 11*(1), 46.

CROWE, S., CRESSWELL, K., ROBERTSON, A., HUBY, G., AVERY, A., & SHEIKH, A. (2011). The case study approach. *BMC Medical Research Methodology, 11*(1), 100. doi: 10.1186/1471-2288-11-100.

CUONG, C., DART, P., DUDLEY, N., & HOCKINGS, M. (2018). Building stakeholder awareness and engagement strategy to enhance biosphere reserve performance and sustainability: The case of Kien Giang, Vietnam. *Environ Manage, 62*(5), 877-891. doi: 10.1007/s00267-018-1094-6.

CUONG, C., DART, P., & HOCKINGS, M. (2017a). Biosphere reserves: Attributes for success. *Journal of Environmental Management, 188*, 9-17. doi: 10.1016/j.jenvman.2016.11.069.

CUONG, C. V., DART, P., DUDLEY, N., & HOCKINGS, M. (2017b). Factors influencing successful implementation of biosphere reserves in Vietnam: Challenges, opportunities and lessons learnt. *Environmental Science & Policy, 67*, 16-26. doi: 10.1016/j.envsci.2016.10.002.

DAKOS, V., QUINLAN, A., BAGGIO, J., BENNETT, E., BODIN, Ö., & BURNSILVER, S. (2015). Principle 2: Manage connectivity. In R. Biggs, M. Schlüter, M.L. Schoon (Eds.), *Principles for building resilience: Sustaining ecosystem services in social-ecological systems* (pp. 80-104). Cambridge: Cambridge University.

DALU, M. T. B., MUKHUWANA, O., CUTHBERT, R. N., MARAMBANYIKA, T., GUNTER, A. W., MURUNGWENI, F. M., & DALU, T. (2022). Understanding communities' perceptions, demographics and uses of wetlands in Vhembe biosphere reserve, South Africa. *Wetlands Ecology and Management, 30*(6), 1231-1244. doi: 10.1007/s11273-022-09892-2.

DANIEL, J. (2011). *Sampling essentials: Practical guidelines for making sampling choices*: SAGE Publications.

DEANGELIS, D. L. (1980). Energy flow, nutrient cycling, and ecosystem resilience. *Ecology, 61*(4), 764-771. doi: 10.2307/1936746.

DELLINGER, A. B. (2005). Validity and the review of literature. *Research in the Schools, 12*(2), 41-54.

DEMIRIS, G., OLIVER, D. P., & WASHINGTON, K. T. (2019). Chapter 3 - Defining and analyzing the problem. In G. Demiris, D. P. Oliver, & K. T. Washington (Eds.), *Behavioral intervention research in hospice and palliative care* (pp. 27-39). London: Academic Press.

DIALLO, I. (2018). *Geopolitics of French in Francophone sub-Saharan Africa*: Cambridge Scholars Publishing.

DOBSON, A., HOPCRAFT, G., MDUMA, S., OGUTU, J. O., FRYXELL, J., ANDERSON, T. M., ARCHIBALD, S., LEHMANN, C., POOLE, J., CARO, T., MULDER, M. B., HOLT, R. D., BERGER, J., RUBENSTEIN, D. I., KAHUMBU, P., CHIDUMAYO, E. N., MILNER-GULLAND, E. J., SCHLUTER, D., OTTO, S., BALMFORD, A., WILCOVE, D., PIMM, S., VELDMAN, J. W., OLFF, H., NOSS, R., HOLDO, R., BEALE, C., HEMPSON, G., KIWANGO, Y., LINDENMAYER, D., BOND, W., RITCHIE, M., & SINCLAIR, A. R. E. (2022). Savannas are vital but overlooked carbon sinks. *Science, 375*(6579), 392-392. doi: 10.1126/science.abn4482.

DONGES, J. F., LUCHT, W., CORNELL, S. E., HEITZIG, J., BARFUSS, W., LADE, S. J., & SCHLÜTER, M. (2021). Taxonomies for structuring models for world–earth systems analysis of the Anthropocene: Subsystems, their interactions and social–ecological feedback loops. *Earth Syst. Dynam., 12*(4), 1115-1137. doi: 10.5194/esd-12-1115-2021.

DOSIO, A., MENTASCHI, L., FISCHER, E. M., & WYSER, K. (2018). Extreme heat waves under 1.5 °C and 2 °C global warming. *Environmental Research Letters, 13*(5), 054006-054006. doi: 10.1088/1748-9326/AAB827.

DROST, E. A. (2011). Validity and reliability in social science research. *Education Research and perspectives, 38*(1), 105-123.

DUBE, K., & NHAMO, G. (2020). Evidence and impact of climate change on South African National Parks. Potential implications for tourism in the Kruger National Park. *Environmental Development, 33*, 100485. doi: 10.1016/j.envdev.2019.100485.

DUDLEY, N. (2008). *Guidelines for applying protected area management categories*: IUCN.

DUDLEY, N., JONAS, H., NELSON, F., PARRISH, J., PYHÄLÄ, A., STOLTON, S., & WATSON, J. E. M. (2018). The essential role of other effective area-based conservation measures in achieving big bold conservation targets. *Global Ecology and Conservation, 15*. doi: 10.1016/j.gecco.2018.e00424.

DUDLEY, N., PARRISH, J. D., REDFORD, K. H., & STOLTON, S. (2010). The revised IUCN protected area management categories: The debate and ways forward. *Oryx, 44*(4), 485-490. doi: 10.1017/S0030605310000566.

DUDLEY, N., & STOLTON, S. (2008). *Defining protected areas: An international conference in Almeria, Spain, May 2007*: IUCN.

DUDLEY, N., & STOLTON, S. (2022). *Best practice in delivering the 30x30 target: protected areas and other effective area-based conservation measures*. https://www.nature.org/content/dam/tnc/nature/en/documents/TNC_UKDEFRA_30x30_BestPractices_Report.pdf: Cambridge.

DUEDAHL, P. (2016). *A history of UNESCO: Global actions and impacts*: Palgrave Macmillan UK.

DUNE, D. (2022). Analysis: Africa's unreported extreme weather in 2022 and climate change. *Carbon Brief.* Retrieved from https://www.preventionweb.net/news/analysis-africas-unreported-extreme-weather-2022-and-climate-change.

EAKIN, H., RUEDA, X., & MAHANTI, A. (2017). Transforming governance in telecoupled food systems. *Ecology and Society, 22*(4). doi: 10.5751/ES-09831-220432.

EDWARDS, J. E. (1997). *How to conduct organizational surveys: A step-by-step guide*: SAGE Publications.

ELLIOTT, H. F. I. (1974). *Second World Conference on National Parks: Proceedings of a conference sponsored & organized by the national parks [and others]*: International Union for Conservation of Nature and Natural Resources.

ELLISON, D., MORRIS, C. E., LOCATELLI, B., SHEIL, D., COHEN, J., MURDIYARSO, D., GUTIERREZ, V., NOORDWIJK, M. V., CREED, I. F., POKORNY, J., GAVEAU, D., SPRACKLEN, D. V., TOBELLA, A. B., ILSTEDT, U., TEULING, A. J., GEBREHIWOT, S. G., SANDS, D. C., MUYS, B., VERBIST, B., SPRINGGAY, E., SUGANDI, Y., & SULLIVAN, C. A. (2017). Trees, forests and water: Cool insights for a hot world. *Global Environmental Change, 43*, 51-61. doi: 10.1016/j.gloenvcha.2017.01.002.

ENGELBAUER, M. (2023). *Global assessment of recent UNESCO Biosphere Reserve quality enhancement strategies and interlinkages with other UNESCO labels*: Würzburg University Press.

EPANDA, M. A., FOTSING, A. J. M., BACHA, T., FRYNTA, D., LENS, L., TCHOUAMO, I. R., & JEF, D. (2019). Linking local people's perception of wildlife and conservation to livelihood and poaching alleviation: A case study of the Dja biosphere reserve, Cameroon. *Acta Oecologica, 97*, 42-48. doi: 10.1016/j.actao.2019.04.006.

ESRI (Cartographer). (2022). African base map. Retrieved from https://www.arcgis.com/apps/mapviewer/index.html?webmap=d582a9e953c44c09bb998c7d9b66f8d4.

EUROPEAN ENVIRONMENT AGENCY. (1999). *Environmental indicators: Typology and overview*: European Environment Agency.

EVERARD, M. (2001). Taking a systems-oriented view of phosphorus enrichment in fresh waters. *Freshwat. Forum 15: 35-54.*

EVERARD, M. (2021). *Ecosystem services: Key issues*: Taylor & Francis.

FALKNER, R. (2016). The Paris Agreement and the new logic of international climate politics. *International Affairs, 92*(5), 1107-1125. doi: 10.1111/1468-2346.12708.

FARAHBAKHSH, I., BAUCH, C. T., & ANAND, M. (2022). Modelling coupled human–environment complexity for the future of the biosphere: Strengths, gaps and promising directions. *Philosophical Transactions of the Royal Society B: Biological Sciences, 377*(1857), 20210382. doi: 10.1098/rstb.2021.0382.

FATH, B. D., DEAN, C. A., & KATZMAIR, H. (2015). Navigating the adaptive cycle: An approach to managing the resilience of social systems. *Ecology and Society, 20*(2). doi: 10.5751/ES-07467-200224.

FAZEY, I., SCHÄPKE ET AL. (2020). Transforming knowledge systems for life on earth: Visions of future systems and how to get there. *Energy Research & Social Science, 70*, 101724. doi: 10.1016/j.erss.2020.101724.

FEELEY, A., FEELEY, I., MERGHANI, K., & SHEEHAN, E. (2021). A pilot study to evaluate the face & construct validity of an orthopaedic virtual reality simulator. *Injury, 52*(7), 1715-1720.

FERNÁNDEZ-FERNÁNDEZ, C. (2021). *Awareness in logic and epistemology: A conceptual schema and logical study of the underlying main epistemic concepts*: Springer International Publishing.

FILHO, W. L., OGUGE, N., AYAL, D., ADELEKE, L., & DA SILVA, I. (2021). *African handbook of climate change adaptation*: Springer International Publishing.

FILHO, W. L., POCIOVALISTEANU, D. M., & AL-AMIN, A. Q. (2016). *Sustainable economic development: Green economy and green growth*: Springer International Publishing.

FINCHAM, J. E. (2008). Response rates and responsiveness for surveys, standards, and the journal. *Am J Pharm Educ, 72*(2), 43. doi: 10.5688/aj720243.

FINK, A., & LITWIN, M. S. (1995). *How to measure survey reliability and validity*: SAGE Publications.

FLEMING, J. (2013). *The Callendar effect: The life and work of Guy Stewart Callendar (1898-1964)*: American Meteorological Society.

FLYNN, B. B., SAKAKIBARA, S., SCHROEDER, R. G., BATES, K. A., & FLYNN, E. J. (1990). Empirical research methods in operations management. *Journal of Operations Management, 9*(2), 250-284. doi: 10.1016/0272-6963(90)90098-X.

FOLKE, C. (2006). Resilience: The emergence of a perspective for social–ecological systems analyses. *Global Environmental Change, 16*(3), 253-267. doi: 10.1016/j.gloenvcha.2006.04.002.

FOLKE, C. (2016). Resilience (republished). *Ecology and Society, 21*(4). doi: 10.5751/ES-09088-210444.

FOLKE, C., BIGGS, R., NORSTRÖM, A. V., REYERS, B., & ROCKSTRÖM, J. (2016). Social-ecological resilience and biosphere-based sustainability science. *Ecology and Society, 21*(3). doi: 10.5751/ES-08748-210341.

FORRESTER, J. W. (1961). *Industrial dynamics*: M.I.T. Press.

FORRESTER, J. W. (1999). *System dynamics: The foundation under systems thinking*: Sloan School of Management.

FORSTER, P. M., SMITH, C. J., WALSH, T., LAMB, W. F., LAMBOLL, R., HAUSER, M., RIBES, A., ROSEN, D., GILLETT, N., & PALMER, M. D. (2023). Indicators of global climate change 2022: Annual update of large-scale indicators of the state of the climate system and human influence. *Earth System Science Data, 15*(6), 2295-2327. doi: 10.5194/essd-15-2295-2023.

FOURIER, J. (2013). Mémoire sur les températures du globe terrestre et des espaces planétaires. In J. G. Darboux (Ed.), *Oeuvres de fourier: Publiées par les soins de gaston darboux* (Vol. 2, pp. 95-126). Cambridge: Cambridge University Press.

FRANKLIN, J. F. (1977). The biosphere reserve program in the United States: A program has been developed to select key sites for environmental research and monitoring. *Science, 195*(4275), 262-267.

FRIELER, K., MEINSHAUSEN, M., GOLLY, A., MENGEL, M., LEBEK, K., DONNER, S. D., & HOEGH-GULDBERG, O. (2012). Limiting global warming to 2 °C is unlikely to save most coral reefs. *Nature Climate Change, 3*(2), 165-170. doi: 10.1038/nclimate1674.

GAAN, N. (2008). *Climate change and international politics*: Kalpaz.

GALLOPÍN, G. C. (2006). Linkages between vulnerability, resilience, and adaptive capacity. *Global Environmental Change, 16*(3), 293-303. doi: 10.1016/j.gloenvcha.2006.02.004.

GANAPATHY, D. (2021). *Media and climate change: Making sense of press narratives*: Taylor & Francis.

GARRICK, J. (1999). Doubting the philosophical assumptions of interpretive research. *International journal of qualitative studies in education, 12*(2), 147-156. doi: 10.1080/095183999236222.

GBÉNAHOU, H. B. M., & SOKEMAWU, K. (2020). Governance of the Pendjari biosphere reserve in view of climate variability. In J. Runge, A. Guézéré, & L. Kankpénandja (Eds.), *Natural resources, socio-ecological sensitivity and climate change in the volta-oti basin, west Africa*. London: CRC Press.

GERMAN NATIONAL UNESCO COMMISSION. (2015). *Management manual for UNESCO biosphere reserves in Africa: Field version*: Deutsche UNESCO-Kommission.

GIORGIO, A. (2003). *Reconciling biodiversity preservation and local sustainable development: The role of the MAB biosphere reserves in the Southeastern European countries* (Doctoral thesis), Università degli studi di Padova.

GISSIBL, B., HÖHLER, S., & KUPPER, P. (2012). *Towards a global history of national parks*: Berghahn Books.

GIVEN, L. M. (2008). *The Sage Encyclopedia of Qualitative Research Methods*: SAGE Publications.

GLERUM, A., ATASOY, B., & BIERLAIRE, M. (2014). Using semi-open questions to integrate perceptions in choice models. *Journal of Choice Modelling, 10*, 11-33. doi: 10.1016/j.jocm.2013.12.001.

GORDON, T., & HELMER, O. (1964). *Report on a long-range forecasting study*: Rand Corporation.

GORDON, T. J. (1994). *The Delphi method (Vol. 3)*: In Futures Research Methodology.

GOURITZ CLUSTER. (2014). Jobs for carbon. Retrieved from https://gouritz.com/project/jobs-for-carbon/.

GREKOUSIS, G. (2020). *Spatial analysis methods and practice: Describe – explore – explain through gis*: Cambridge University Press.

GUNDERSON, L. H. (2000). Ecological resilience in theory and application. *Annual Review of Ecology and Systematics, 31*, 425-439.

GUNDERSON, L. H., & HOLLING, C. S. (2002). *Panarchy: Understanding transformations in human and natural systems*: Island Press.

GUPTA, J., TERMEER, C., KLOSTERMANN, J., MEIJERINK, S., VAN DEN BRINK, M., JONG, P., NOOTEBOOM, S., & BERGSMA, E. (2010). The adaptive capacity wheel: A method to assess the inherent characteristics of institutions to enable the adaptive capacity of society. *Environmental Science & Policy, 13*(6), 459-471. doi: 10.1016/j.envsci.2010.05.006.

HALLEGATTE, S., & ROZENBERG, J. (2017). Climate change through a poverty lens. *Nature Climate Change 2017, 7*(4), 250-256. doi: 10.1038/nclimate3253.

HARDIN, G. (1968). The tragedy of the commons. *Science, 162*(3859), 1243-1248.

HARKIOLAKIS, N. (2017). *Quantitative research methods: From theory to publication*: CreateSpace Independent Publishing Platform.

HENLEY, B. J., & KING, A. D. (2017). Trajectories toward the 1.5°C Paris target: Modulation by the interdecadal Pacific oscillation. *Geophysical Research Letters, 44*(9), 4256-4262. doi: 10.1002/2017GL073480.

HENNINK, M., HUTTER, I., & BAILEY, A. (2010). *Qualitative research methods*: SAGE Publications.

HEYMANN, M., & DAHAN DALMEDICO, A. (2019). Epistemology and politics in earth system modeling: Historical perspectives. *Journal of Advances in Modeling Earth Systems, 11*(5), 1139-1152. doi: 10.1029/2018MS001526.

HINKEL, J., BOTS, P. W. G., & SCHLÜTER, M. (2014). Enhancing the Ostrom social-ecological system framework through formalization. *Ecology and Society, 19*(3).

HOCKING, B. (1985). Words and deeds: Why America left UNESCO. *The World Today, 41*(4), 75-78.

HOEGH-GULDBERG, O., JACOB, D., TAYLOR, M., BINDI, M., BROWN, S., CAMILLONI, I., DIEDHIOU, A., DJALANTE, R., EBI, K.L., ENGELBRECHT, F., GUIOT, J., HIJIOKA, Y., MEHROTRA, S., PAYNE, A., SENEVIRATNE, S.I., THOMAS, A., WARREN, R., & ZHOU, G. (2018). Impacts of 1.5°C Global Warming on Natural and Human Systems. In V. Masson-Delmotte, P. Zhai, H.-O. Pörtner, D. Roberts, J. Skea, P.R. Shukla, A. Pirani, W. Moufouma-Okia, C. Péan, R. Pidcock, S. Connors, J.B.R. Matthews, Y. Chen, X. Zhou, M.I.Gomis, E. Lonnoy, T.Maycock, M.Tignor, & T. Waterfield (Eds.), Global Warming of 1.5°C. *An IPCC Special Report on the impacts of global warming of 1.5°C above pre-industrial levels and related global greenhouse gas emission pathways, in the context of strengthening the global response to the threat of climate change, sustainable development, and efforts to eradicate poverty* (pp.175-311): IPCC.

HOLLING, C. S. (1973). Resilience and stability of ecological systems. *Annual Review of Ecology and Systematics, 4*(1), 1-23. doi: 10.1146/annurev.es.04.110173.000245.

HOLLING, C. S. (1986). The resilience of terrestrial ecosystems; local surprise and global change. In W. C. Clark & R. E. Munn (Eds.), *Sustainable development of the biosphere* (pp. 292-317). Cambridge: Cambridge University Press.

HOLLING, C. S. (1996). *Engineering resilience versus ecological resilience*. Washington: National Academy of Science.

HOLLING, C. S. (2001). Understanding the complexity of economic, ecological, and social systems. *Ecosystems, 4*(5), 390-405.

HOLLING, C. S., & GUNDERSON, L. H. (2002). Resilience and adaptive cycles. In L. H. Gunderson & C. S. Holling (Eds.), *Panarchy: Understanding transformations in human and natural systems* (pp. 25-62). Washington: Island Press.

HULUKA, A. T., & WONDIMAGEGNHU, B. A. (2019). Determinants of household dietary diversity in the Yayo biosphere reserve of Ethiopia: An empirical analysis using sustainable livelihood framework. *Cogent Food & Agriculture, 5*(1), 1690829. doi: 10.1080/23311932.2019.1690829.

IMF. (2020). *Adapting to climate change in sub-Saharan Africa*: IMF.

IMF. (2023). *GDP per capita, current prices*: IMF.

INTERNATIONAL LEGAL MATERIALS. (1989). Vienna convention for the protection of the ozone layer and its Montreal protocol: Helsinki declaration of participating states. *International Legal Materials, 28*(5), 1335-1336. doi: 10.1017/S0020782900022816.

INTRONA, L., KAVANAGH, D., KELLY, S., ORLIKOWSKI, W., & SCOTT, S. (2016). *Beyond interpretivism? New encounters with technology and organization: IFIP wg 8.2 working conference on information systems and organizations, IS&O 2016, Dublin, Ireland, December 9-10, 2016, proceedings*: Springer International Publishing.

IPCC. (1990). *IPCC first assessment report*: WMO Geneva.

IPCC. (1995). *Climate change 1995: IPCC second assessment report; a report of the intergovernmental panel on climate change*: IPCC.

IPCC. (2001). *Climate change 2001 ; IPCC third assessment report*: IPCC.

IPCC. (2013). Summary for policymakers. In T. F. Stocker, D. Qin, G. K. Plattner, M. Tignor, S. K. Allen, J. Boschung, A. Nauels, Y. Xia, V. Bex & P. M. Midgley (Eds.), *Climate change 2013: The physical science basis. Contribution of working group I to the fifth assessment report of the IPCC*. Cambridge: Cambridge University Press.

IPCC (2014). *Climate change 2014: Synthesis report. Contribution of working groups i, ii and iii to the fifth assessment report of the intergovernmental panel on climate change*: IPCC.

IPCC. (2018). An IPCC special report on the impacts of global warming of 1.5°C above pre-industrial levels and related global greenhouse gas emission pathways, in the context of strengthening the global response to the threat of climate change, sustainable development, and efforts to eradicate poverty. In V. Masson-Delmotte, P. Zhai, H.-O. Pörtner, D. Roberts, J. Skea, P.R. Shukla, A. Pirani, W. Moufouma-Okia, C. Péan, R. Pidcock, S. Connors, J.B.R. Matthews, Y. Chen, X. Zhou, M.I. Gomis, E. Lonnoy, T. Maycock, M. Tignor, & T. Waterfield (Eds.), *Global warming of 1.5°C*: IPCC.

IPCC. (2022). *Sixth assessment report fact sheet: Africa*: IPCC.

IPCC. (2014). *IPCC fifth assessment report—synthesis report*: IPCC.

Isaac, N. (2021, 27 October). COP26: Why Africa's voice must be heard. *Science Nigeria*. Retrieved from https://sciencenigeria.com/cop26-why-africas-voice-must-be-heard/.

Ison, R., & Straw, E. (2020). *The hidden power of systems thinking: Governance in a climate emergency*: Taylor & Francis.

IUCN. (1978). *Categories, objectives and criteria for protected areas*: IUCN.

IUCN. (1994). *Guideline for protected area management categories*: IUCN.

IUCN. (2017). *Guidelines for recognising and reporting other effective area-based conservation methods*: IUCN.

Jahn, A. (2018). Reduced probability of ice-free summers for 1.5°C compared to 2°C warming. *Nature Climate Change, 8*(5), 409-413. doi: 10.1038/S41558-018-0127-8.

James Madison University. (2023). *Empirical research papers: An overview*. Retrieved from https://www.jmu.edu/uwc/_files/link-library/empirical/empirical_research_article_overview.pdf.

JEVREJEVA, S., JACKSON, L. P., GRINSTED, A., LINCKE, D., & MARZEION, B. (2018). Flood damage costs under the sea level rise with warming of 1.5 °C and 2 °C. *Environmental Research Letters, 13*(7), 074014-074014. doi: 10.1088/1748-9326/aacc76.

JOB, H. (2014). Bevölkerungswachstum und Schutzgebiete – eine afrikanische Herausforderung. *Geographische Rundschau,* 66 (10), 44-47.

JOB, H. (2022). Eine Idee erobert die Welt: 150 Jahre Nationalparks. *Geographische Rundschau*, 74 (3), 4-9.

JOB, H., BECKEN, S., & LANE, B. (2017). Protected Areas in a neoliberal world and the role of tourism in supporting conservation and sustainable development: an assessment of strategic planning, zoning, impact monitoring, and tourism management at natural World Heritage Sites. *Journal of Sustainable Tourism,* 25 (12), 1697-1718. doi: 10.1080/09669582.2017.1377432.

JOB, H., ENGELBAUER, M., & ENGELS, B. (2019). Das Portfolio deutscher Biosphärenreservate im Lichte der Sustainable Development Goals. *Raumforschung und Raumordnung,* 77 (1), 57-79. doi: 10.2478/rara-2019-0005.et

JOB, H., MAJEWSKI, L., BITTLINGMAIER, S., ENGELBAUER, M., & WOLTERING, M. (2023). *Regionalökonomische Effekte des Tourismus in Biosphärenreservaten Deutschlands*: BfN-Schriften 667.

JOB, H., MEYER, C., CORONADO, O., KOBLAR, S., LANER, P., OMIZZOLO, A., PLASSMANN, G., RIEDLER, W., VESELY, P., & SCHINDELEGGER, A. (2022). Open spaces in the European Alps—GIS-based analysis and implications for spatial planning from a transnational perspective. *Land, 11*, 1605. doi: 10.3390/land11091605.

JOB, H., & WEIZENEGGER, S. (2006). Actor-Oriented Management Strategies for Protected Areas. In: Natural Heritage, Ecotourism and Sustainable Development. In Job, H., & Li, J. (Eds.), *Münchner Studien zur Sozial- und Wirtschaftsgeographie* 45, 103-114.

JOFFE, H., & YARDLEY, L. (2003). Chapter four: Content and thematic analysis. In D. Marks & l. Yardley (Eds.), *Research methods for clinical and health psychology* (pp. 56-68). London: Sage.

JOHNSON, R. B., & CHRISTENSEN, L. (2019). *Educational research: Quantitative, qualitative, and mixed approaches*: Sage publications.

JUSSIM, L. (2001). Self-fulfilling prophecies. In N. J. Smelser & P. B. Baltes (Eds.), *International Encyclopedia of the Social & Behavioral Sciences* (pp. 13830-13833). Oxford: Pergamon.

KABIR, S. M. (2016). Methods of data collection. In *Basic guidelines for research: An introductory approach for all disciplines*: Mohammad Javed Rahim.

KARLING, H. M. (2001). *Global climate change*: Nova Publishers.

KARROS, D. J. (1997). Statistical methodology: Reliability and validity assessment in study design, part b. *Academic Emergency Medicine, 4*(2), 144-147.

KEELING, R. F. (2008). Recording Earth's vital signs. *Science, 319*(5871), 1771-1772.

KELKER, H., YODER, K., MUSEY, P., HARRIS, M., JOHNSON, O., SARMIENTO, E., VYAS, P., HENDERSON, B., ADAMS, Z., & WELCH, J. L. (2020). Longitudinal prospective study of emergency medicine provider wellness across ten academic and community hospitals during the initial surge of the Covid-19 pandemic. *BMC Emerg Med, 21(1):36*. Doi: 10.21203/rs.3.rs-87786/v.

KELLER, C. F. (1999). Comment: Human contribution to climate change increasingly clear. *EOS Transactions, 80*, 368-372. doi: 10.1029/EO080i033p00368-01.

KI-MOON, B. (2013). The Millennium Development Goals Report 2013. *United Nations Pubns, 365*, 366.

KIMARIO, F. F., BOTHA, N., KISINGO, A. W., & JOB, H. (2020). Theory and practice of conservancies: Evidence from wildlife management areas in Tanzania. *Erdkunde*, 117-143. doi: 10.3112/erdkunde.2020.02.03.

KITSIKOPOULOS, H. (2015). *Innovation and technological diffusion: An economic history of early steam engines*: Taylor & Francis.

KOTSCHY, K., BIGGS, R., DAW, T., FOLKE, C., & WEST, P. (2015). Principle 1 – maintain diversity and redundancy. In *Principles for Building Resilience: Sustaining Ecosystem Services in Social-Ecological Systems* (pp. 50-79). Cambridge: Cambridge University Press.

KRAUS, F., MERLIN, C., & JOB, H. (2014). Biosphere reserves and their contribution to sustainable development. *Zeitschrift für Wirtschaftsgeographie, 58*(1), 164-180. Doi: 10.1515/zfw.2014.0011.

KUDAKWASHE, M., ZVIKONYAUKWA, J., & MUDZENGI, C. (2023). Exploring challenges for the sustainable management of wildlife resources in Matusadonha National Park and its hinterland. *Journal of Sustainable Development in Africa, 25*(1), 21-36.

KUH, K. F. (2018). The law of climate change mitigation: An overview. In D. A. Dellasala & M. I. Goldstein (Eds.), *Encyclopedia of the anthropocene* (pp. 505-510). Oxford: Elsevier.

KULLMANN, F., MARKEWITZ, P., STOLTEN, D., & ROBINIUS, M. (2021). Combining the worlds of energy systems and material flow analysis: A review. *Energy, Sustainability and Society, 11*(1), 13. doi: 10.1186/s13705-021-00289-2.

KUMAR, A. (2022). *Ecosystem-based adaptation: Approaches to sustainable management of aquatic resources*: Elsevier Science.

KUPIKA, O. L., GANDIWA, E., & NHAMO, G. (2019a). Green economy initiatives in the face of climate change: Experiences from the Middle Zambezi Biosphere Reserve, Zimbabwe. *Environment, Development and Sustainability, 21*(5), 2507-2533. doi: 10.1007/s10668-018-0146-7.

KUPIKA, O. L., GANDIWA, E., NHAMO, G., & KATIVU, S. (2019b). Local ecological knowledge on climate change and ecosystem-based adaptation strategies promote resilience in the Middle Zambezi Biosphere Reserve, Zimbabwe. *Scientifica (Cairo), 2019*, 3069254. doi: 10.1155/2019/3069254.

KUŠOVÁ, D., TĚŠITEL, J., MATĚJKA, K., & BARTOŠ, M. (2008). Biosphere reserves—an attempt to form sustainable landscapes: A case study of three biosphere reserves in the Czech Republic. *Landscape and urban planning, 84*(1), 38-51. doi: 10.1016/j.landurbplan.2007.06.006.

LAUNGANI, P. (2007). *Understanding cross-cultural psychology: Eastern and western perspectives*: SAGE Publications.

LE, Q. B., & FEITOSA, F. (2012). *Comparison of two common empirical methods to model land-use choices in a multi-agent system simulation of landscape transition: Implication for a hybrid approach*: International Congress on Environmental Modelling and Software.

LE TREUT, H., SOMERVILLE, R., CUBASCH, U., DING, Y., MAURITZEN, C., MOKSSIT , A., PETERSON, T., & PRATHER, M. (2007). Historical overview of climate change. In S. Solomon, D. Qin, M. Manning, Z. Chen, M. Marquis, K.B. Averyt, M. Tignor and & H. L. Miller (Eds.), *Climate change 2007: The physical science basis. Contribution of working group I to the fourth assessment report of the IPCC*. Cambridge: Cambridge University Press.

LEMKE, P., REN, J. F., ALLEY, R., ALLISON, I., CARRASCO, J., FLATO, G., FUJII, Y., KASER, G., MOTE, P., THOMAS, R., & ZHANG, T. (2007). IPCC, 2007. Climate change 2007. Synthesis report. Contribution of working groups I, II & III to the fourth assessment report of the Intergovernmental Panel on Climate Change. Geneva. In https://www.ipcc.ch/report/ar4/syr/.

LENNAN, M., & MORGERA, E. (2022). The Glasgow Climate Conference (COP26). *The International Journal of Marine and Coastal Law, 37*(1), 137-151. doi: 10.1163/15718085-bja10083.

LEWIN, K. (1951). *Field theory in social science: Selected theoretical papers (edited by Dorwin Cartwright.)*: Harpers.

LEWIS, S. L., & MASLIN, M. A. (2015). Defining the Anthropocene. *Nature, 519*(7542), 171-180. doi: 10.1038/nature14258.

LINSTONE, H., & TUROFF, M. (1975). *The Delphi method: Techniques and applications (Vol. 18)*:Wiley.

LO, V., & JANG, N. (2022). The global biodiversity framework's "30x30" target: Catchy slogan or effective conservation goal? Retrieved from https://www.iisd.org/articles/insight/global-biodiversity-framework-30x30-target.

LOWE, D. B. (2016). *The chemistry book: From gunpowder to graphene, 250 milestones in the history of chemistry*: Union Square & Company.

MAB. (1996). Biosphere reserves: The Seville strategy and the statutory framework of the World Network. Retrieved from http://www.mab.cas.cn/ryswqjh/swqbhq/201411/W020141113678526165131.pdf.

MAB. (2011). *International co-ordinating council of the man and the biosphere (MAB) programme, twenty-third session, Dresden Radebeul (Germany): Final report:* UNESCO.

MACHLIS, G. E., FORCE, J. E., & BURCH, W. R. (1997). The human ecosystem Part I: The human ecosystem as an organizing concept in ecosystem management. *Society & Natural Resources, 10*(4), 347-367. doi: 10.1080/08941929709381034.

MAIBACH, E., LEISEROWITZ, A., COBB, S., SHANK, M., COBB, K., & GULLEDGE, J. (2012). The legacy of climategate: Undermining or revitalizing climate science and policy? *Wiley interdisciplinary reviews: Climate Change, 3*, 289-295. doi: 10.1002/wcc.168.

MANDAL, J., & PARIJA, S. C. (2014). Informed consent and research. *Tropical parasitology, 4*(2), 78. doi: 10.4103/2229-5070.138533.

MANLEY, G. (1974). Central England Temperatures: Monthly means 1659 to 1973. *Quarterly Journal of the Royal Meteorological Society, 100*, 389-405. doi: 10.1002/qj.49710042511.

MARUSTER, L. (2013). *Qualitative research methods*: SAGE Publications.

MASOOD, E., & TOLLEFSON, J. (2021). 'COP26 hasn't solved the problem': Scientists react to un climate deal. *Nature 2016 534:7609*(599), 355-356. doi: 10.1038/d41586-021-03431-4.

MAY, A. (2020). *Politics and climate change: A history*: Andy May Petrophysicist.

MAYER, B. (2022). *International law obligations on climate change mitigation*: OUP Oxford.

MAYER, M., BRENNER, L., SCHAUSS, B., STADLER, C., ARNEGGER, J., &JOB, H. (2018). The nexus between governance and the economic impact of whale-watching. The case of coastal lagoons in the El Vizcaíno Biosphere Reserve, Baja California, Mexico. *Ocean and Coastal Management,* 162, 46-59. doi: 10.1016/j.ocecoaman.2018.04.016.

McCormick, J. (1995). *The global environmental movement*: John Wiley & Sons.

McDonald, M. (2002). *Systematic assessment of learning outcomes: Developing multiple-choice exams*: Jones and Bartlett Publishers.

McGinnis, M. D., & Ostrom, E. (2014). Social-ecological system framework: Initial changes and continuing challenges. *Ecology and Society, 19*(2). doi: 10.5751/ES-06387-190230.

Meadows, D. (2008). *Thinking in systems: International bestseller*: Chelsea Green Publishing.

Mehren, R., & Rempfler, A. (2022). Assessing systems thinking in geography. In T. Bourke, R. Mills, & R. Lane (Eds.), *Assessment in geographical education: An international perspective* (pp. 31–54). Cham: Springer.

Meijering, J. V., & Tobi, H. (2018). The effects of feeding back experts' own initial ratings in Delphi studies: A randomized trial. *International Journal of Forecasting, 34*(2), 216-224. doi: 10.1016/j.ijforecast.2017.11.010.

Mello, P. A., Ide, T., Andersson, M. A., Harkness, S. K., Swinkels, M., Giordono, L., Boudet, H., Gard-Murray, A., Castillo-Ortiz, P., & Brockhaus, M. (2021). *Qualitative comparative analysis: An introduction to research design and application*: Georgetown University Press.

Menard, S. (2007). *Handbook of longitudinal research: Design, measurement, and analysis*: Elsevier Science.

Merton, R. (1948). The self-fulfilling prophecy. *The Antioch Review, 8*, 193-210. doi: 10.2307/4609267.

Messick, S. (1995). Validity of psychological assessment: Validation of inferences from persons' responses and performances as scientific inquiry into score meaning. *American psychologist, 50*(9), 741-749. doi: 10.1037/0003-066X.50.9.741.

Middleton, W. E. K. (1966). *A history of the thermometer and its use in meteorology*: Johns Hopkins University Press.

Millar, E., Mitiku, F., Hiruy, H. N., & Barden-O'Fallon, J. (2023). Differences by age group in gender-related attitudes in Oromia, Ethiopia: Findings from a cross-sectional study of women of reproductive age. *African Journal of Reproductive Health, 27*(5), 36-45.

Mitchard, E. T. A. (2018). The tropical forest carbon cycle and climate change. *Nature, 559*(7715), 527-534. doi: 10.1038/s41586-018-0300-2.

Mitsch, W. J., Bernal, B., Nahlik, A. M., Mander, Ü., Zhang, L., Anderson, C. J., Jørgensen, S. E., & Brix, H. (2013). Wetlands, carbon, and climate change. *Landscape ecology, 28*, 583-597. doi: 10.1007/s10980-012-9758-8.

MÖLLER, L. (2011). *For life, for the future: Biosphere reserves and climate change: A collection of good practice case studies*: German Commission for UNESCO.

MORIARTY, P., & HONNERY, D. (2018). Energy policy and economics under climate change. *AIMS Energy, 6*(2), 272-290. doi: 10.3934/energy.2018.2.272.

MORSE, J. M. (2010). Simultaneous and sequential qualitative mixed method designs. *Qualitative inquiry, 16*(6), 483-491. doi: 10.1177/1077800410364741.

MORSE, J. M. (2016). *Mixed method design: Principles and procedures*: Taylor & Francis.

MORSE, S., & MCNAMARA, N. (2013). *Sustainable livelihood approach: A critique of theory and practice*: Springer Netherlands.

MURAIRWA, S. (2015). Voluntary sampling design. *4*, 185-200.

NADERIFAR, M., GOLI, H., & GHALJAEI, F. (2017). Snowball sampling: A purposeful method of sampling in qualitative research. *Strides in Development of Medical Education, 14*(3). doi: 10.5812/sdme.67670.

NAOHIRO, G., ULHASANAH, N., KAMAHARA, H., HASANUDIN, U., TACHIBANA, R., & FUJIE, K. (2015). Material and energy flow analysis. In J. Dewulf, S. De Meester, A Rodrigo, & R. A. F. Alvarenga (Eds.), *Sustainability assessment of renewables-based products* (pp. 125-140). Chichester: John Wiley & Sons.

NATIONAL RESEARCH COUNCIL. (2011). *Climate stabilization targets: Emissions, concentrations, and impacts over decades to millennia*: National Academies Press.

NATTRASS, B., & ALTOMARE, M. (1999). *The natural step for business: Wealth, ecology and the evolutionary corporation*: New Society Publishers.

NEWSOME, D., MOORE, S. A., & DOWLING, R. K. (2013). *Natural area tourism: Ecology, impacts and management*: Channel View Publications.

NIXON, S. J. C. (2016). *The invention of the automobile: Karl Benz and Gottlieb Daimler*: Edizioni Savine.

NOAA. (2023). *Trends in atmospheric carbon dioxide*: NOAA.

NORDHAUS, W. D. (2007). A review of the "Stern review on the economics of climate change". *Journal of Economic Literature, 45*(3), 686-702.

O'LEARY-KELLY, S. W., & VOKURKA, R. J. (1998). The empirical assessment of construct validity. *Journal of Operations Management, 16*(4), 387-405. doi: 10.1016/S0272-6963(98)00020-5.

OECD. (2018). *Mainstreamin biodiversity for sustainable development*: OECD.

OFLAZOGLU, S. (2017). *Qualitative versus quantitative research*: IntechOpen.

OLADOYE, A., IGE, P., BAURWA, N., ONILUDE, Q., & ANIMASHAUN, Z. (2020). Slenderness coefficient models for tree species in Omo Biosphere Reserve, Southwestern Nigeria. *Tropical Plant Research, 7*(3), 609-618. doi: 10.22271/tpr.2020.v7.i3.075.

Oliveira, B. G., Liboni, L. B., Cezarino, L. O., Stefanelli, N. O., & Miura, I. K. (2020). Industry 4.0 in systems thinking: From a narrow to a broad spectrum. *Systems Research and Behavioral Science, 37*(4), 593-606. doi: 10.1002/sres.2703.

Olsson, P., Folke, C., & Berkes, F. (2004). Adaptive comanagement for building resilience in social–ecological systems. *Environmental Management, 34*(1), 75-90. doi: 10.1007/s00267-003-0101-7.

Ostrom, E. (2009). A general framework for analyzing sustainability of social-ecological systems. *Science, 325*(5939), 419-422. doi: 10.1126/science.1172133.

Ostrom, E. (2010). Polycentric systems for coping with collective action and global environmental change. *Global Environmental Change, 20*(4), 550-557. doi: 10.1016/j.gloenvcha.2010.07.004.

Oxford English Dictionary. (2023). *Oxford English Dictionary*: Oxford University Press.

Padmavathy, R. D., Anand, K., Kekare, S., & Cheepi, P. (2023). *Research methodology methods, tools & techniques*: Academic Guru Publishing House.

Pahl-Wostl, C., Holtz, G., Kastens, B., & Knieper, C. (2010). Analyzing complex water governance regimes: The management and transition framework. *Environmental Science & Policy, 13*(7), 571-581. doi: 10.1016/j.envsci.2010.08.006.

Pantić, M., Colic, N., & Milijić, S. (2021). Golija-Studenica Biosphere Reserve (Serbia) as a driver of change. *Eco.mont (Journal on Protected Mountain Areas Research), 13*, 58-69. doi: 10.1553/eco.mont-13-sis58.

Parker, C. (2020). *Ford Model T enthusiasts' manual: 1908 to 1927 - an insight into the design, production and ownership of the car that introduced motoring to the masses*: Haynes Publishing Group P.L.C.

Parkinson, C. L., & Comiso, J. C. (2013). On the 2012 record low Arctic Sea ice cover: Combined impact of preconditioning and an August storm. *Geophysical Research Letters, 40*(7), 1356-1361. doi: 10.1002/grl.50349.

Partelow, S. (2018). A review of the social-ecological systems framework: Applications, methods, modifications, and challenges. *Ecology and Society, 23*. doi: 10.5751/ES-10594-230436.

Patten, M. L., & Galvan, M. C. (2019). *Proposing empirical research: A guide to the fundamentals*: Taylor & Francis.

Paül, V., Vila-Lage, R., & Trillo-Santamaría, J.-M. (2022). "The No1 country"? A critical investigation of the booming designation of biosphere reserves in Spain. *Landscape and urban planning, 222*, 104375. doi: 10.1016/j.landurbplan.2022.104375.

Pereira, L. F. (2023). Business research methods: systematic literature review. *International Journal of Management Concepts and Philosophy*, 16(2), 145-159. doi: 10.1504/IJMCP.2023.130029.

PETROSILLO, I., ARETANO, R., & ZURLINI, G. (2015). *Socioecological systems, reference module in Earth systems and environmental sciences:* Elsevier.

PHAM, L. T. M. (2018). *Qualitative approach to research a review of advantages and disadvantages of three paradigms: Positivism, interpretivism and critical inquiry*: University of Adelaide.

PHILLIPS, A. (2004). A short history of the international system of protected area management categories. *Parks, 14*(3), 4-14.

PLASS, G. N. (1956). The carbon dioxide theory of climatic change. *Tellus, 8*(2), 140-154. doi: 10.3402/tellusa.v8i2.8969.

Plieninger, T., Woltering, M., & Job, H. (2016). Implementierung des Ökosystemleistungs-Ansatzes in deutschen Biosphärenreservaten. *Raumforschung und Raumordnung, 74*(5), 541-554. doi: 10.1007/s13147-016-0438-z.

POOL-STANVLIET, R. (2013). A history of the UNESCO Man and the Biosphere programme in South Africa. *South African Journal of Science, 109*, 1-6.

PREUSS, P. (2013). *Data-driven decision making and dynamic planning*: Taylor & Francis.

PRICE, M. F. (1996). People in biosphere reserves: An evolving concept. *Society & Natural Resources, 9*(6), 645-654. doi: 10.1080/08941929609381002.

PRICE, M. F., PARK, J. J., & BOUAMRANE, M. (2010). Reporting progress on internationally designated sites: The periodic review of biosphere reserves. *Environmental science & policy, 13*(6), 549-557. doi: 10.1016/j.envsci.2010.06.005.

PROVENZALE, A., & KILGARRIFF, A. (2023). *History of climate change: From the Earth's origins to the Anthropocene*: Polity Press.

PRYSHLAKIVSKY, J., & SEARCY, C. (2021). Life cycle assessment as a decision-making tool: Practitioner and managerial considerations. *Journal of Cleaner Production, 309*, 127344. doi: 10.1016/j.jclepro.2021.127344.

PÜTZ, M., & JOB, H. (2016). Governance und Regionalentwicklung in Großschutzgebieten der Schweiz und Österreichs. *Raumforschung und Raumordnung, 74*(5), 569-583. doi: 10.1007/s13147-016-0451-2.

QUEVEDO, J. M. D., LUKMAN, K. M., ULUMUDDIN, Y. I., UCHIYAMA, Y., & KOHSAKA, R. (2023). Applying the DPSIR framework to qualitatively assess the globally important mangrove ecosystems of Indonesia: A review towards evidence-based policymaking approaches. *Marine Policy, 147*, 105354. doi: 10.1016/j.marpol.2022.105354.

RAMAGE, M., & SHIPP, K. (2020). *Systems thinkers*: Springer London.

RAMOS-QUINTANA, F., SOTELO-NAVA, H., SALDARRIAGA-NOREÑA, H., & TOVAR-SÁNCHEZ, E. (2019). Assessing the environmental quality resulting from damages to human-nature interactions caused by population increase: A systems thinking approach. *Sustainability, 11*(7), 1957. doi: 10.3390/su11071957.

RANDLE, J. M., & STROINK, M. L. (2018). The development and initial validation of the paradigm of systems thinking. *Systems Research and Behavioral Science, 35*(6), 645-657. doi: 10.1002/sres.2508.

RANNIKKO, P. (1996). Local environmental conflicts and the change in environmental consciousness. *Acta sociologica, 39*(1), 57-72.

RASMUSSEN, D. J., BITTERMANN, K., BUCHANAN, M. K., KULP, S., STRAUSS, B. H., KOPP, R. E., & OPPENHEIMER, M. (2018). Extreme sea level implications of 1.5 °C, 2.0 °C, and 2.5 °C temperature stabilization targets in the 21st and 22nd centuries. *Environmental Research Letters, 13*(3), 034040-034040. doi: 10.1088/1748-9326/aaac87.

RATZLAFF, E. D. (1969). *Applications of engineering systems analysis to the human social-ecological system*: University of California, Davis.

REED, M. G. (2016). Conservation (in)action: Renewing the relevance of UNESCO Biosphere Reserves. *Conservation Letters, 9*(6), 448-456. doi: 10.1111/conl.12275.

REED, M. G., & MASSIE, M. M. (2013). Embracing ecological learning and social learning UNESCO Biosphere Reserves as exemplars of changing conservation practices. *Conservation and Society, 11*(4), 391-405.

REED, M. G., & PRICE, M. F. (2019). *UNESCO Biosphere Reserves: Supporting biocultural diversity, sustainability and society*: Taylor & Francis.

RESCHIWATI, R., & SOEPRIJADI, S. (2018). Path analysis: The determinant of student learning outcomes in accounting introduction courses (study at private higher education in Jakarta). *Global and Stochastic Analysis, 5*(5).

RITTBERGER, V., ZANGL, B., KRUCK, A., & DIJKSTRA, H. (2019). *International organization*: Bloomsbury Publishing.

ROBINSON, D. E. (2010). *Climategate debunked: Big brother, mainstream media, cover-ups*: CreateSpace Independent Publishing Platform.

RODRÍGUEZ-RODRÍGUEZ, D., SÁNCHEZ-ESPINOSA, A., & MALAK, D. A. (2021). Potential contribution of OECMs to international area-based conservation targets in a biodiversity rich country, Spain. *Journal for Nature Conservation, 62*, 126019. doi: 10.1016/j.jnc.2021.126019.

ROSEN, A. (2015). The wrong solution at the right time: The failure of the Kyoto Protocol on climate change. *Politics & Policy, 43*. doi: 10.1111/polp.12105.

ROSENTHAL, R. (2012). Self-fulfilling prophecy. In V. S. Ramachandran (Ed.), *Encyclopedia of human behavior (second edition)* (pp. 328-335). San Diego: Academic Press.

ROWE, G., & WRIGHT, G. (1999). The delphi technique as a forecasting tool: Issues and analysis. *International Journal of Forecasting, 15*(4), 353-375. doi: 10.1016/S0169-2070(99)00018-7.

RYAN, V. D., AGNITSCH, K. A., ZHAO, L., & MULLICK, R. (2005). Making sense of voluntary participation: A theoretical synthesis. *Rural Sociology, 70*(3), 287-313. doi: 10.1526/0036011054831198.

SACHER, P., & MAYER, M. P. D. (2019). *Szenarien der Waldentwicklung in Bayern bis 2045 und 2075 – Ergebnisse einer Delphi-Studie mit Ausgewählten Waldexperten*: Institut für Geographie und Geologie der Universität Greifswald.

SACHS, J. D. (2012). From millennium development goals to sustainable development goals. *The Lancet, 379*(9832), 2206-2211. doi: 10.1016/S0140-6736(12)60685-0.

SADLER, G. R., LEE, H. C., LIM, R. S. H., & FULLERTON, J. (2010). Recruitment of hard-to-reach population subgroups via adaptations of the snowball sampling strategy. *Nursing & health sciences, 12*(3), 369-374. doi: 10.1111/j.1442-2018.2010.00541.x.

SALMON, P. M., STANTON, N. A., WALKER, G. H., HULME, A., GOODE, N., THOMPSON, J., & READ, G. J. M. (2022). *Handbook of systems thinking methods*: CRC Press.

SANGHA, K. K., PREECE, L., VILLARREAL-ROSAS, J., KEGAMBA, J. J., PAUDYAL, K., WARMENHOVEN, T., & RAMAKRISHNAN, P. S. (2018). An ecosystem services framework to evaluate indigenous and local peoples' connections with nature. *Ecosystem Services, 31*, 111-125. doi: 10.1016/j.ecoser.2018.03.017.

SAYWARD, A. L. (2017). *The United Nations in International History*: Bloomsbury Publishing.

SCHILLING, J., HERTIG, E., TRAMBLAY, Y., & SCHEFFRAN, J. (2020). Climate change vulnerability, water resources and social implications in North Africa. *Regional Environmental Change, 20*, 1-12. doi: 10.1007/s10113-020-01597-7.

SCHLEUSSNER, C. F., LISSNER, T. K., FISCHER, E. M., WOHLAND, J., PERRETTE, M., GOLLY, A., ROGELJ, J., CHILDERS, K., SCHEWE, J., FRIELER, K., MENGEL, M., HARE, W., & SCHAEFFER, M. (2016). Differential climate impacts for policy-relevant limits to global warming: The case of 1.5°C and 2°C. *Earth System Dynamics, 7*(2), 327-351. doi:10.5194/ESD-7-327-2016.

SCHLIEP, R., & STOLL-KLEEMANN, S. (2010). Assessing governance of biosphere reserves in Central Europe. *Land Use Policy, 27*(3), 917-927. doi: 10.1016/j.landusepol.2009.12.005.

SCHLÜTER, M., HAIDER, L. J., LADE, S. J., LINDKVIST, E., MARTIN, R., ORACH, K., WIJERMANS, N., & FOLKE, C. (2019). Capturing emergent phenomena in social-ecological systems an analytical framework. *Ecology and Society, 24*(3). doi: 10.5751/ES-11012-240311.

SCHOLZ, R., & BINDER, C. (2004). *Principles of human-environment systems research.* Paper presented at the IEMSs 2004, June 14-17, 2004. http://hdl.handle.net/20.500.11850/85916.

SCHOONENBOOM, J., & JOHNSON, R. B. (2017). How to construct a mixed methods research design. *Kölner Z Soz Sozpsychol, 69*(2), 107-131. doi: 10.1007/s11577-017-0454-1.

SCHROGL, K. U., GIANNOPAPA, C., & ANTONI, N. (2021). *A research agenda for space policy*: Edward Elgar Publishing.

SCHULTZ, L., DUIT, A., & FOLKE, C. (2011). Participation, adaptive co-management, and management performance in the world network of biosphere reserves. *World Development, 39*(4), 662-671. doi: 10.1016/j.worlddev.2010.09.014.

SCHULTZ, L., & LUNDHOLM, C. (2010). Learning for resilience? Exploring learning opportunities in biosphere reserves. *Environmental Education Research, 16*(5-6), 645-663. doi: 10.1080/13504622.2010.505442.

SCHULTZ, L., WEST, S., BOURKE, A. J., D'ARMENGOL, L., TORRENTS, P., HARDARDOTTIR, H., JANSSON, A., & ROLDÁN, A. M. (2018). Learning to live with social-ecological complexity: An interpretive analysis of learning in 11 UNESCO Biosphere Reserves. *Global Environmental Change, 50*, 75-87. doi: 10.1016/j.gloenvcha.2018.03.001.

SCHUMPETER, J. A. (2013). *Capitalism, socialism and democracy*: Taylor & Francis.

SCHUTT, R. K. (2009). *Investigating the social world: The process and practice of research*: Pine Forge Press.

SCHWARTZ-SHEA, P., & YANOW, D. (2012). *Interpretive research design: Concepts and processes*: Routledge.

SEDGWICK, P. (2013). Questionnaire surveys: Sources of bias. *BMJ, 347.*

SENGE, P. M. (1990). *The fifth discipline: The art and practice of the learning organization*: Doubleday.

SENGE, P. M., SMITH, B. J., ROBERTS, C., & KLEINER, A. (1994). *The fifth discipline fieldbook: Strategies and tools for building a learning organization*: Crown.

SERRAT, O. (2017). The Sustainable Livelihoods Approach. In O. Serrat (Ed.) *Knowledge Solutions* (pp.21-26). Singapore: Springer.

SHOWSTACK, R. (2013). Carbon dioxide tops 400 ppm at Mauna Loa, Hawaii. *Eos, Transactions American Geophysical Union, 94*(21), 192-192. doi: 10.1002/2013EO210004.

SIMON, J. (2017). *The art of empirical investigation*: Taylor & Francis.

SIMONSEN, S. H., BIGGS, R. O., SCHLÜTER, M., SCHOON, M., BOHENSKY, E., CUNDILL, G., DAKOS, V., DAW, T., KOTSCHY, K., LEITCH, A., QUINLAN, A., PETERSON, G., & MOBERG, F. (2014). *Applying resilience thinking*: Stockholm Resilience Centre.

SINGH, K. (2009). *Quantitative social research methods*: SAGE Publications.

SKIBINS, J. C., POWELL, R. B., & HALLO, J. C. (2013). Charisma and conservation: Charismatic megafauna's influence on safari and zoo tourists' pro-conservation behaviors. *Biodiversity and conservation, 22*, 959-982. doi: 10.1007/s10531-013-0462-z.

SLOCUM, T. A., MCMASTER, R. B., KESSLER, F., & HOWARD, H. H. (2022). *Thematic cartography and geovisualization*: CRC Press, Taylor & Francis Group.

ŚLUSARSKI, M., & JURKIEWICZ, M. (2019). Visualisation of spatial data uncertainty. A case study of a database of topographic objects. *ISPRS International Journal of Geo-Information, 9*(1), 16. doi: 10.3390/ijgi9010016.

SMALL, C., PARNELL, G., POHL, E., GOERGER, S., COTTAM, B., SPECKING, E., & WADE, Z. (2018). *Engineering resilience for complex systems*: Springer.

STATISTA. (2021). Global carbon dioxide emissions per capita 2021, by country. Retrieved from https://www.statista.com/statistics/270508/co2-emissions-per-capita-by-country/.

STIENS, G. (1977). Zur Verwendung von Szenarien in der Raumplanung. *Raumforschung und Raumordnung, 35*(1), 2.

STIENS, G. (2003). *Szenarien zur Raumentwicklung: Raum- und Siedlungsstrukturen Deutschlands 2015/2040*. Bonn-Bad Godesberg: Bundesamt für Bauwesen und Raumordnung.

STOLL-KLEEMANN, S., DE LA VEGA-LEINERT, A., & SCHULTZ, L. (2010). The role of community participation in the effectiveness of UNESCO Biosphere Reserve management: Evidence and reflections from two parallel global surveys. *Environmental Conservation, 37*, 227-238. doi: 10.1017/S037689291000038X.

STOLL-KLEEMANN, S., & JOB, H. (2008). The Relevance of Effective Protected Areas for Biodiversity Conservation: An Introduction. *Gaia*, 17 (S1), 86-90.

STOLL-KLEEMANN, S., & O'RIORDAN, T. (2017). The challenges of the Anthropocene for biosphere reserves. *Parks, 23*, 89-100. doi: 10.2305/IUCN.CH.2017.PARKS-23-1SS-K.en.

STOLL-KLEEMANN, S., & WELP, M. (2008). Participatory and integrated management of biosphere reserves: Lessons from case studies and global survey. *GAIA, 17*(1), 161-168. doi: 10.14512/gaia.17.S1.14.

STOLL-KLEEMANN, S., & WELP, M. (2010). *Stakeholder dialogues in natural resources management: Theory and practice*: Springer Berlin Heidelberg.

STRATTON, S. J. (2023). Population sampling: Probability and non-probability techniques. *Prehospital and Disaster Medicine, 38*(2), 147-148. doi: 10.1017/S1049023X23000304.

STRAUSS, M. E., & SMITH, G. T. (2009). Construct validity: Advances in theory and methodology. *Annual review of clinical psychology, 5*, 1-25. doi: 10.1146/annurev.clinpsy.032408.153639.

STROUP, R. L. (2000). Free riders and collective action revisited. In P. Fishback, G. Libecap, & E. Zajac (Eds.), *Public choice essays in honour of a maverick scholar Gordon Tullock.* Boston: Springer.

STUKAS, A. A., & SNYDER, M. (2022). Self-fulfilling prophecies. In H. S. Friedman (Ed.), *Encyclopedia of mental health* (pp. 92-100). Elsevier.

TAGHIPOORREYNEH, M. (2023). Mixed methods and the Delphi method. In R. J. Tierney, F. Rizvi, & K. Ercikan (Eds.), *International encyclopedia of education (fourth edition)* (pp. 608-614). Oxford: Elsevier.

TALBOT, L. M. (1980). A world conservation strategy. *Journal of the Royal Society of Arts, 128*(5288), 493-510.

TASHAKKORI, A., JOHNSON, R. B., & TEDDLIE, C. (2020). *Foundations of mixed methods research: Integrating quantitative and qualitative approaches in the social and behavioral sciences*: SAGE Publications.

TASHAKKORI, A., & TEDDLIE, C. (1998). *Mixed methodology: Combining qualitative and quantitative approaches*: SAGE Publications.

TAVAKOL, M., & DENNICK, R. (2011). Making sense of Cronbach's Alpha. *International journal of medical education, 2*, 53. doi: 10.5116/ijme.4dfb.8dfd.

TAYLOR, M. (2014). *Global warming and climate change: What Australia knew and buried... Then framed a new reality for the public*: ANU Press.

TAYLOR, P., FAHRIG, L., & WITH, K. (2006). Landscape connectivity: A return to the basics. In K.R. Crooks, & M. Sanjayan (Eds.), *Connectivity conservation* (pp. 29-43). Cambridge: Cambridge University Press.

TAYLOR, S. J., BOGDAN, R., & DEVAULT, M. (2015). *Introduction to qualitative research methods: A guidebook and resource*: Wiley.

TEAGUE, K. A., & GALLICCHIO, N. (2017). *The evolution of meteorology: A look into the past, present, and future of weather forecasting*: Wiley.

THE ROYAL SOCIETY. (2014). *Climate change: Evidence and causes*: National Academies Press.

TIME. (1953, May 25). Invisible blanket. *Time*.

TRI, C., HENS, L., HUNG, N., TRAN, P., & PHUOC, P. (2017). A systematic approach to the dilemma between flood vulnerability and resilience-review and concepts. *Vietnam Journal of Science and Technology, 55*, 620-636. doi: 10.15625/2525-2518/55/5/9195.

TROIAN, A., GOMES, M. C., TIECHER, T., BERBEL, J., & GUTIÉRREZ-MARTÍN, C. (2021). The drivers-pressures-state-impact-response model to structure cause–effect relationships between agriculture and aquatic ecosystems. *Sustainability, 13*(16), 9365. doi: 10.3390/su13169365.

TSIVACOU, I. (1993). *A challenge for systems thinking: The Aegean seminar*: University of the Aegean Press.

TSOMB, E. I. B. T., NSOGA, M. H. I. I. I. N., & BITTING, C. D. (2023). Climate change vulnerability and conflicts in Africa: Evidence from the migrations channel. *Environment, Development and Sustainability*. doi: 10.1007/s10668-023-03417-4.

TUCKER, C. M. (2013). *Developing sustainability indicators for the Kogelberg and Cape West Coast biosphere reserves, South Africa*: Stellenbosch University Press.

TURNER, B. L., KASPERSON, R. E., MATSON, P. A., MCCARTHY, J. J., CORELL, R. W., CHRISTENSEN, L., ECKLEY, N., KASPERSON, J. X., LUERS, A., MARTELLO, M. L., POLSKY, C., PULSIPHER, A., & SCHILLER, A. (2003). A framework for vulnerability analysis in sustainability science. *Proceedings of the National Academy of Sciences, 100*(14), 8074-8079. doi: 10.1073/pnas.1231335100.

TURNER, J. R., & BAKER, R. M. (2019). Complexity theory: An overview with potential applications for the social sciences. *Systems, 7*(1). doi: 10.3390/systems7010004.

TYNDALL, J. (1861). The Bakerian lecture: On the absorption and radiation of heat by gases and vapours, and on the physical connexion of radiation, absorption, and conduction. *Philosophical Transactions of the Royal Society of London, 151*, 1-36. doi: 10.1098/rspl.1860.0021.

UN. (1973). *Report of the United Nations Conference on the Human Environment, Stockholm, 5-16 June 1972*: UN.

UN. (2023). *Causes and effects of climate change*: UN.

UNDP. (2007). *Human development report 2007/2008 - fighting climate change: Human solidarity in a divided world*: UNDP.

UNEP. (2015). *The emissions gap report 2015*: UNEP.

UNEP. (2020). *Six ways nature can protect us from climate change*: UNEP.

UNESCO. (1968). *Intergovernmental conference of experts on the scientific basis for rational use and conservation of the resources of the biosphere (Paris, France, September 4-13, 1968), recommendations*: UNESCO.

UNESCO. (1974). *Final report: Programme on man and the biosphere (MAB), task force on: Criteria and guidelines for the choice and establishment of biosphere reserves; organized jointly by UNESCO and UNEP, Paris, 20-24 May, 1974*: UNESCO.

UNESCO. (1984). *Action plan for biosphere reserves*: UNESCO.

UNESCO. (1995a). *Implementation of 27 c/resolution 2.3, concerning the Seville strategy on biosphere reserves and the statutory framework of the world network of biosphere reserves*: UNESCO.

UNESCO. (1995b). *International conference on biosphere reserves, Seville, Spain, 1995: Final report MAB report series*: UNESCO.

UNESCO. (2000). *Seville +5; international meeting of experts on the implementation of the Seville strategy for biosphere reserves, Pamplona, Spain, 23-27 October 2000*: UNESCO.

UNESCO. (2008). *Madrid action plan for biosphere reserves (2008-2013)*: UNESCO.

UNESCO. (2010a). *Glossary of basic terminology on disaster risk reduction*: UNESCO.

UNESCO. (2010b). *Lessons from biosphere reserves in the Asia-Pacific region, and a way forward: A regional review of biosphere reserves in Asia & the Pacific to achieve sustainable development*: UNESCO.

UNESCO. (2014). *Man and the biosphere (MAB) programme strategy for the period 2015-2025*: UNESCO.

UNESCO. (2016). *Lima action plan for UNESCO's Man and the Biosphere (MAB) programme and its world network of biosphere reserves (2016-2025)*: UNESCO.

UNESCO. (2017). *A new roadmap for the Man and the Biosphere (MAB) programme and its world network of biosphere reserves*: UNESCO.

UNESCO. (2018). *UNESCO MAB programme leaflet*: UNESCO.

UNESCO. (2020). *Statutory framework of the world network of biosphere reserves*: UNESCO.

UNESCO. (2022). *The role of UNESCO biosphere reserves in the implementation of the convention on biological diversity's post-2020 global biodiversity framework: Policy brief*: UNESCO.

UNESCO. (2023). *Biosphere reserves in Africa*: UNESCO.

UNFCCC. (2016). *Adoption of the Paris Agreement*: UNFCCC.

UNFCCC. (2020). *Climate change is an increasing threat to Africa*: UNFCCC.

UNFCCC. (2021). *The Glasgow Climate Pact – key outcomes from COP26*: UNFCCC.

UNFCCC. (2023). Five key takeaways from COP27. Retrieved from https://unfccc.int/process-and-meetings/conferences/sharm-el-sheikh-climate-change-conference-november-2022/five-key-takeaways-from-cop27.

UNIVERSTY OF NOTRE DAME. (2021). *ND-GAIN Country Index*. Retrieved from: https://gain.nd.edu/our-work/country-index/rankings/.

US PRESIDENT'S SCIENCE ADVISORY COMMITTEE. (1965). *Restoring the quality of our environment: Report*: White House.

VAN ASSCHE, K., VALENTINOV, V., & VERSCHRAEGEN, G. (2019). Ludwig von Bertalanffy and his enduring relevance: Celebrating 50 years general system theory. *Systems Research and Behavioral Science, 36*(3), 251-254. doi: 10.1002/sres.2589.

VAN KERKHOFF, L. A.-O., MUNERA, C., DUDLEY, N., GUEVARA, O., WYBORN, C., FIGUEROA, C., DUNLOP, M., HOYOS, M. A., CASTIBLANCO, J., & BECERRA, L. Towards future-oriented conservation: Managing protected areas in an era of climate change. *Ambio, 48*(7), 699-713. doi: 10.1007/s13280-018-1121-0.

VELDERS, G. J. M., ANDERSEN, S. O., DANIEL, J. S., FAHEY, D. W., & MCFARLAND, M. (2007). The importance of the Montreal Protocol in protecting climate. *Proceedings of the National Academy of Sciences, 104*(12), 4814-4819. doi: 10.1073/pnas.0610328104.

VERMA, J. P. (2019). *Statistics and research methods in psychology with Excel*: Springer Nature Singapore.

VIRAPONGSE, A., BROOKS, S., METCALF, E. C., ZEDALIS, M., GOSZ, J., KLISKEY, A., & ALESSA, L. (2016). A social-ecological systems approach for environmental management. *Journal of Environmental Management, 178*, 83-91. doi: 10.1016/j.jenvman.2016.02.028.

VOGEL, C., ZWOLINSKY, S., GRIFFITHS, C., HOBBS, M., HENDERSON, E., & WILKINS, E. (2019). A Delphi study to build consensus on the definition and use of big data in obesity research. *International Journal of Obesity, 43*(12), 2573-2586. doi: 10.1038/s41366-018-0313-9.

VOGLER, J., & BRETHERTON, C. (2006). The European Union as a protagonist to the United States on climate change. *International Studies Perspectives, 7*(1), 1-22.

VON BERTALANFFY, L. (1969). *General system theory: Foundations, development, applications*: George Braziller.

VON WEIZSÄCKER, E. U., TARASOFSKY, R. G., OBERTHÜR, S., & OTT, H. E. (1999). *The Kyoto Protocol: International climate policy for the 21st century*: Springer Berlin Heidelberg.

WALKER, B., CARPENTER, S., ANDERIES, J., ABEL, N., CUMMING, G., JANSSEN, M., LEBEL, L., NORBERG, J., PETERSON, G., & PRITCHARD, R. (2002). Resilience management in social-ecological systems: A working hypothesis for a participatory approach. *Ecology and Society, 6*(1), 14.

WALKER, B., HOLLING, C.S., CARPENTER, S., & KINZIG, A. (2003). Resilience, adaptability and transformability in social-ecological systems. *Ecol. Soc., 9*. doi: 10.5751/ES-00650-090205.

WALKER, B., & SALT, D. (2006). *Resilience thinking: Sustaining ecosystems and people in a changing world*: Island Press.

WALLNER, A., BAUER, N., & HUNZIKER, M. (2007). Perceptions and evaluations of biosphere reserves by local residents in Switzerland and Ukraine. *Landscape and urban planning, 83*(2), 104-114. doi: 10.1016/j.landurbplan.2007.03.006.

WAMSLER, C. (2017). Stakeholder involvement in strategic adaptation planning: Transdisciplinarity and co-production at stake? *Environmental Science & Policy, 75*, 148-157. doi: 10.1016/j.envsci.2017.03.016.

WAN ROSELY, W. I. H., & VOULVOULIS, N. (2022). Systems thinking for the sustainability transformation of urban water systems. *Critical Reviews in Environmental Science and Technology*, 1-21. doi: 10.1080/10643389.2022.2131338.

WANG, X., & CHENG, Z. (2020). Cross-sectional studies: Strengths, weaknesses, and recommendations. *Chest, 158*(1, Supplement), S65-S71. doi: 10.1016/j.chest.2020.03.012.

WARREN, R., PRICE, J., GRAHAM, E., FORSTENHAEUSLER, N., & VAN DER WAL, J. (2018). The projected effect on insects, vertebrates, and plants of limiting global warming to 1.5°C rather than 2°C. *Science, 360*(6390), 791-795. doi: 10.1126/SCIENCE.AAR3646.

WEART, S. R. (2008). *The discovery of global warming: Revised and expanded edition*: Harvard University Press.

WHITE, L., & EDWARDS, A. E. (2000). *Conservation research in the African rain forests: A technical handbook*: Wildlife Conservation Society.

WHITE, R. M., SCHMOOK, B., CALMÉ, S., GIORDANO, A. J., HAUSSER, Y., KIMMEL, L., LECUYER, L., LUCHERINI, M., MÉNDEZ-MEDINA, C., & PEÑA-MONDRAGÓN, J. L. (2023). Facilitating biodiversity conservation through partnerships to achieve transformative outcomes. *Conservation Biology, 37*(3), 14057. doi: 10.1111/cobi.14057.

WHO. (2014). Quantitative risk assessment of the effects of climate change on selected causes of death, 2030s and 2050s. Retreived from https://www.who.int/publications-detail-redirect/9789241507691.

WIENER, N. (1948). *Cybernetics: Or Control and Communication in the Animal and the Machine*: J. Wiley.

WILES, R., CROW, G., HEATH, S., & CHARLES, V. (2008). Anonymity and confidentiality. *International Journal of Social Research Methodology, 11*(5) 417-428. doi: 10.1080/13645570701622231.

WILSON, A. H. (2017). *Encyclopedia of local history*: Rowman & Littlefield Publishers.

WINKLER, K. (2019). The implementation of the conceptual shift in conservation: Pathways of three German UNESCO Biosphere Reserves. *15*, 173-180. doi: 10.1080/26395916.2019.1617351.

WOOD, M. J., & KERR, J. C. (2006). *Basic steps in panning nursing research: From question to proposal*. Boston: Jones and Bartlett Publishers.

WYBORN, C. (2011). Landscape scale ecological connectivity: Australian survey and rehearsals. *Pacific Conservation Biology, 17*(2), 121-131.

XU, Y., & RAMANATHAN, V. (2017). Well below 2°C: Mitigation strategies for avoiding dangerous to catastrophic climate changes. *Proceedings of the National Academy of Sciences, 114*(39), 10315-10323. doi: 10.1073/PNAS.1618481114.

YOOPETCH, C., KONGARCHAPATARA, B., & NIMSAI, S. (2023). Tourism forecasting using the delphi method and implications for sustainable tourism development. *Sustainability, 15*(1), 126. doi: 10.3390/su15010126.

YOUSUF, M. I. (2007). Using experts' opinions through Delphi technique. *Practical Assessment, Research, and Evaluation, 12*(4). doi: 10.7275/rrph-t210.

ZALASIEWICZ, J., WATERS, C. N., WILLIAMS, M., & SUMMERHAYES, C. P. (2019). *The Anthropocene as a geological time unit: A guide to the scientific evidence and current debate*: Cambridge University Press.

ZAMANZADEH, V., GHAHRAMANIAN, A., RASSOULI, M., ABBASZADEH, A., ALAVI-MAJD, H., & NIKANFAR, A. R. (2015). Design and implementation content validity study: Development of an instrument for measuring patient-centered communication. *J Caring Sci, 4*(2), 165-178. doi: 10.15171/jcs.2015.017.

ZHANG, Y., & AYYUB, B. M. (2021). Chapter 2 - Temperature extremes in a changing climate. In A. Fares (Ed.), *Climate change and extreme events* (pp. 9-23): Elsevier.

ZILLMAN, J. (2009). A history of climate activities. *WMO Bulletin, 58*(3), 141-150.

Appendix A

First questionnaire

BRClimateChangeProjects → base 07.12.2023, 14:28

Page 01

"Climate change prevention and adaptation strategies in UNESCO biosphere reserves"

Dear expert, thank you for agreeing to share your knowledge and expertise in this questionnaire.

This research is towards a doctorate in Geography at the Julius-Maximilians-Universität Würzburg, Germany. Its overall aim is to determine how UNESCO biosphere reserves can position themselves in such a way that they can optimally counter climate change within their current institutional structure. This questionnaire's goal is to gather information on your organisation's contribution to climate change mitigation and adaptation within the UNESCO biosphere reserve network. Three climate change scenarios will be presented and your opinion of your organisation's contribution towards these measures, and the role of the UNESCO biosphere reserve network, would be appreciated. This is a Delphi type questionnaire, meaning the results will be gathered and presented to you within six months, after which you will be asked to complete a follow-up questionnaire. After I finished my research I will gladly provide you with a free copy of the thesis describing the results.

Please note: as this questionnaire will be sent out to a variety of stakeholders in the UNESCO BR network, not all questions may apply to you. Please skip these by writing "not applicable " or „ n/a".

1. If you would like to receive the results and participate in the second round of questionnaires, please provide your email address.

Email:

2. Please be so kind and also add your telephone number. This is only needed if I have further queries regarding your statements.

Phone/SMS:

Page 02
S1p1

3. With which Regional and Thematic Networks within the World Network of Biosphere Reserves are you most associated.

Please check all relevant categories in order of involvement

Regional network

- ☐ Africa
- ☐ Arab states
- ☐ Asia and the pacific
- ☐ Europe and North America
- ☐ Latin America and the Caribbean, Portugal and Spain

Thematic network

- ☐ Drylands
- ☐ Mangroves
- ☐ Marine, coastal and Island Areas
- ☐ Mountains
- ☐ Savannah
- ☐ Tropical forest
- ☐ Wetlands

4. Please indicate at which level you are most involved with the UNESCO MAB biosphere reserve network.

- ☐ International – please state how
- ☐ National – name the country
- ☐ Local – name the biosphere reserve
- ☐ Other

5. Which stakeholder group(s) do you primarily represent in the UNESCO MAB biosphere reserve network?
Please indicate all relevant categories.

- ☐ International Coordinating Council
- ☐ Thematic/regional network administration
- ☐ Thematic/regional network membership
- ☐ National committee administration
- ☐ National committee membership
- ☐ Biosphere reserve coordinator/chairperson
- ☐ Formal conservation area management
- ☐ Government organisation – please indicate national, provincial or municipal level
- ☐ Biosphere reserve board member / director – please state portfolio
- ☐ Closely affiliated company to the biosphere reserve – please state economic sector
- ☐ Research – please state field
- ☐ Consultation/advisory – please state field
- ☐ Other:

6. If different from the biosphere reserve, what is the name of the organisation that you represent within the UNESCO biosphere reserve network?

Page 03
s2p2

1. Climate change scenario: Business as usual

The UNESCO biosphere reserve network supports the goal of the Glasgow Pact (building on the Paris Agreement) to limit global average temperature increases by encouraging UNESCO biosphere reserves to become priority sites for climate change monitoring, research, adaptation, and mitigation. If current greenhouse gas emissions continue, the global average temperature could increase to approximately 4°C from pre-industrial levels by 2100, surpassing the 2°C target by 2050. This would lead to extreme heat events, already fast-rising sea levels, declining biodiversity, polar ice melt, the destruction of coral reefs, fishery decrease, increasing poverty, societal health deterioration, and food shortages.

7. In or near which UNESCO biosphere reserve(s) are climate change related projects of your organisation located?

8. In general, since when has your organisation been involved with climate change related projects?

9. Since 2000, how many climate change related projects have been pursued by your organisation within the UNESCO biosphere reserve network?

10. Please name the climate change related projects you are currently busy with / have executed in the past.

If applicable, please attach the abstract or the website link to any climate change related projects that your organisation is involved in within your UNESCO biosphere reserve network.

11. Do these projects relate to any of the following categories?

Please check all relevant categories

- ☐ Forestry that mitigates climate change including reduced deforestation and reforestation
- ☐ Adaptation to climate change in forestry
- ☐ Mitigation to climate change in agriculture and/or husbandry
- ☐ Adaptation to climate change in agriculture and/or husbandry (including the revitalization of traditional farming knowledge).
- ☐ Adaptation through improved mosaics of natural and cultivated spaces
- ☐ Balancing conflicting interests and newly competing forms of land use
- ☐ Eliminating subsidies harmful to biodiversity
- ☐ Avoiding degradation of permafrost areas
- ☐ Inland wetland restoration
- ☐ Coastal zone adaptation (sea-level rise, storm surges, hurricanes, groundwater salinity)
- ☐ Adaptation to climate change in river/flood management
- ☐ Sustainable fishing
- ☐ Rebuilding marine megafauna
- ☐ Restoring degraded (semi-) arid ecosystems
- ☐ Adaptation to increasing water scarcity/droughts/desertification
- ☐ Using fire and bush removal to combat woody plant encroachment
- ☐ Rewilding with large terrestrial mammals
- ☐ Increasing the formal area under protection
- ☐ Enhancing landscape connectivity through area-based conservation and/or 'other effective area-based conservation measures' that achieve conservation goals outside formal protected areas (e.g., biodiversity corridors).
- ☐ Adaptation to shifting vegetation
- ☐ Adaptation to new species (migratory and/or invasive)
- ☐ Adaptation to specifically threatened species
- ☐ Mainstreaming biodiversity
- ☐ Reducing pollution (all sources)
- ☐ Implementing emission reduction trading mechanisms
- ☐ Low emission public transport
- ☐ Promotion of non-conflictual biofuels (e.g., agricultural residues or grass cuttings)
- ☐ Improved stoves for simple solid biomass
- ☐ Mitigating the residual impact of development (e.g., mining, urban development, agricultural expansion, green agriculture initiatives, sustainable forestry, hunting, anti-poaching programmes).
- ☐ Improving energy efficiency in buildings and cities
- ☐ Enhancing biodiversity conservation in transformed ecosystems
- ☐ Sustainable consumption patterns
- ☐ Low-carbon macro-economic approaches (regional marketing and circular economic flows)
- ☐ Adaptation in the tourism industry (e.g.,low impact tourism)
- ☐ Creating income from local tourism-related carbon credit mechanisms

12. Are these projects funded on an international, national/federal, state/provincial, or local level?

- ☐ International
- ☐ Federal/national
- ☐ State/provincial
- ☐ Local
- ☐ All of the above

13. In which biosphere reserve zone(s) are climate change adaptation or mitigation projects being implemented?

Please select the relevant option(s)

- ☐ Core (e.g., formal conservation areas)
- ☐ Buffer (e.g., private reserves, conservancies or OECMs)
- ☐ Transition (e.g., urban, or industrial areas)
- ☐ All of the above
- ☐ Don't know

14. Is the UNESCO biosphere reserve directly involved in any of these projects?

Please indicate the correct role(s).

- ☐ Leading role
- ☐ Partner role
- ☐ Coordinating role
- ☐ Observer role
- ☐ No role
- ☐ Other

15. Does the organisation you represent implement the following:

Please indicate all relevant categories with regards to climate change

- ☐ Climate change planning priorities in the participatory process
- ☐ Adapting planning, implementing and evaluation processes of the management plan and its frequency
- ☐ Involving new stakeholders including, traditional and indigenous communities from outside the biosphere reserve
- ☐ Integrating policies of several government entities/departments/agencies
- ☐ Local climate change scenarios/predictions
- ☐ Long term climate change monitoring
- ☐ New interdisciplinary research
- ☐ Generating new knowledge by combining results across projects
- ☐ Involving political/economical stakeholders into research design
- ☐ Innovative climate change education for children and students
- ☐ Creating public awareness/understanding for climate change
- ☐ Mutual learning among stakeholders about climate change
- ☐ Sensitizing decision-makers and donors about climate change
- ☐ Other

16. Climate change is an important issue in the management plan of the biosphere reserve

- ☐ Fully correct
- ☐ Mostly correct
- ☐ Partially correct
- ☐ Not correct
- ☐ Don't know

17. The biosphere reserve administration (and/or the government entity in charge) has formulated a specific climate change strategy/action plan for the region.

- ☐ Fully correct
- ☐ Mostly correct
- ☐ Partially correct
- ☐ Not correct
- ☐ Don't know

18. The biosphere reserve has positioned itself specifically concerning climate change – it is a „hotspot" and/or climate change is its main focus.

- ☐ Fully correct
- ☐ Mostly correct
- ☐ Partially correct
- ☐ Not correct
- ☐ Don't know

19. The biosphere reserve has a specific budget for climate change actions.

- ☐ Fully correct
- ☐ Mostly correct
- ☐ Partially correct
- ☐ Not correct
- ☐ Don't know

20. Were the Dresden Declaration, Aichi Biodiversity Targets, Paris Agreement, and/or the Glasgow Pact contributing to the decision to pursue these projects?

- ☐ Yes
- ☐ No
- ☐ Don't know

21. Do you know of relevant projects implemented by other individuals or organisations in the biosphere reserve?

Please name these projects and individuals or organisations responsible for them.

Page 04
S3p3

2. Climate change scenario: Middling targets

The Glasgow Pact aims to limit the global average temperature increase to 2ºC by 2100, which would mean that by 2050 the global average temperature increase must remain below 2ºC. This goal is supported by the UNESCO biosphere reserve network and could be accomplished if climate change action is taken. A 2ºC temperature increase has the following estimated impacts:

- Hot days in mid-latitude regions would be 4°C warmer with 37% of the global population experiencing extreme heatwaves.
- A median sea-level rise of 56-63 cm by 2100 with increasing incidences of coastal floods.
- 13% change in terrestrial ecosystems, resulting in the loss of more than half the climatically determined habitats of 23-38% of plants, 16-37% of insects, and 1-14% of vertebrates.
- The Arctic would become ice-free every 10 years, which would impact weather systems and ocean currents.
- 99% loss of coral reefs
- A 2°C increase will result in a decline of 3 million tonnes of harvested fish annually.
- A 2°C global temperature rise is expected to reduce wheat crops by 12% and rice yields by 20%.

22. What would you say is the probability that the global increase in temperature would remain below 2ºC by 2050?

	0	10	20	30	40	50	60	70	80	90	100
Probability of staying below 2ºC by 2050											

23. Has there been research conducted on the amount of carbon dioxide (CO2) emitted from your biosphere reserve?

☐ Yes

☐ No

24. In your opinion, is the organisation you represent in the UNESCO biosphere reserve doing as much as possible to work towards limiting global temperature increases to below 2ºC?

Please rank their contribution, with 1 being the least they can do and 10 being the most.

1 2 3 4 5 6 7 8 9 10

25. How active is the organisation(s) in the biosphere reserve's core zone (independent from the BR) in implementing climate change mitigation and adaptation projects?

☐ Very active

☐ Moderately active

☐ Not very active

☐ Inactive

☐ Don't know

26. How active is the organisation(s) in the biosphere reserve's buffer zone (independent from the BR) in implementing climate change mitigation and adaptation projects?

- ☐ Very active
- ☐ Moderately active
- ☐ Not very active
- ☐ Inactive
- ☐ Don't know

27. How active is the organisation(s) in the biosphere reserve's transition zone (independent from the BR) in implementing climate change mitigation and adaptation projects?

- ☐ Very active
- ☐ Moderately active
- ☐ Not very active
- ☐ Inactive
- ☐ Don't know

28. Do you think the UNESCO biosphere reserve status and its global network have a substantial impact on climate change adaptation and mitigation?

Please elaborate.

Page 05
S4p4

3. Climate change scenario: Ideal targets

The ideal aim of the Glasgow Pact would be to limit the global temperature increase to 1.5° C by 2100. The temperature increase trend would therefore need to plateau between 1° C and 1.5° C by 2050.

29. What would you say is the probability that the global increase in temperature would plateau between 1° C and 1.5° C by 2050?

	0	10	20	30	40	50	60	70	80	90	100
The probability of temperature plateau is between 1° C and 1.5° C.											

30. What is the share of your organisation's climate change related projects to all projects running in or around your UNESCO biosphere reserve network?

Please indicate the approximate percentage.

	0	10	20	30	40	50	60	70	80	90	100
Share of climate change projects in %											

31. What are the factors that determine whether a climate change project will be pursued by your organisation?

Please list all factors separately.

32. Do you have any further suggestions on how the UNESCO biosphere reserve network can help limit the global temperature increase to below 1.5° C by 2050?

Page 06

33. Were you aware of the details of any of the climate change scenarios mentioned above before doing this questionnaire?

- ☐ Yes
- ☐ Partly
- ☐ No

34. In your opinion, in which climate change scenario does your organisation's efforts fit?

- ☐ Scenario 1: Business as usual
- ☐ Scenario 2: Middling target
- ☐ Scenario 3: Ideal target

35. Which scenario do you think is most likely to happen?

- ☐ Scenario 1, Business as usual: temperature increase of more than 2ºC by 2050.
- ☐ Scenario 2, Middling target: temperature increase remains just below 2ºC by 2050.
- ☐ Scenario 3, Ideal target: temperature plateau between 1º C and 1.5º C by 2050.

Last Page

Thank you for completing this questionnaire!

Your answers were transmitted, you may close the browser window or tab now.

Appendix B

Second questionnaire

delphi_survey_2 → base 07.12.2023, 17:14

Page 01

“Climate change prevention and adaptation strategies in UNESCO biosphere reserves”

Dear expert, thank you for agreeing to share your knowledge and expertise in this questionnaire.

This is the second round of a Delphi questionnaire which you completed six months ago. This research is towards a doctorate in Geography at the Julius-Maximilians-Universität Würzburg, Germany. Its overall aim is to determine how UNESCO biosphere reserves can position themselves in such a way that they can optimally counter climate change within their current institutional structure. This questionnaire aims to gather information on the impact of climate change on UNESCO biosphere reserves and your organisation's contribution to climate change mitigation and adaptation. After I finished my research I will gladly provide you with a free copy of the thesis describing the results.

Please note: as this questionnaire will be sent out to a variety of stakeholders in the UNESCO BR network, not all questions may apply to you. Please skip these by writing "not applicable “ or „ n/a".

1. Which country do you represent?

Please select from the following:

[Please choose]

2. Which UNESCO biosphere reserve(s) do you represent or have intimate knowledge of?

Please select from the following:

- ☐ Aïr et Ténéré
- ☐ Amboseli
- ☐ Badiar
- ☐ Basse-Lobaye
- ☐ Belo-sur-Mer Kirindy-Mite
- ☐ Benoué
- ☐ Bia
- ☐ Black River Gorges-Bel Ombre
- ☐ Boloma-Bijagós
- ☐ Boucle du Baoulé
- ☐ Cape West Coast
- ☐ Cape Winelands
- ☐ Chimanimani
- ☐ Comoé
- ☐ Complex W-Arly-Penjari (WAP) Transboundary
- ☐ Complexe Oti-Kéran / Oti-Mandouri
- ☐ Delta du Fleuve Sénégal Transboundary
- ☐ Delta du Saloum
- ☐ Dimonika
- ☐ Dja
- ☐ Doumba-Rey
- ☐ East Usambara
- ☐ Ferlo
- ☐ Fogo
- ☐ Gadabedji
- ☐ Garden Route
- ☐ Gishwati-Mukura Landscape
- ☐ Gombe Masito Ugalla
- ☐ Gouritz Cluster
- ☐ Hadejia Nguru Bade
- ☐ Haut Niger
- ☐ Jozani – Chwaka Bay
- ☐ Kafa
- ☐ Kafue Flats
- ☐ Kiunga
- ☐ Kogelberg
- ☐ Kruger to Canyons
- ☐ Lake Bosomtwe
- ☐ Lake Chilwa Wetland
- ☐ Lake Manyara
- ☐ Lake Tana
- ☐ Littoral de Toliara
- ☐ Lubombo

- ☐ Luki
- ☐ Magaliesberg
- ☐ Maio
- ☐ Majang Forest
- ☐ Malindi Watamu Arabuko Sokoke
- ☐ Mananara Nord
- ☐ Mare aux hippopotames
- ☐ Marico
- ☐ Massif du Ziama
- ☐ Matšeng
- ☐ Middle Zambezi
- ☐ Mono Transboundary
- ☐ Monts Nimba
- ☐ Mount Elgon
- ☐ Mount Kenya – Lewa
- ☐ Mount Kulal
- ☐ Mount Mulanje
- ☐ Mwali
- ☐ Niokolo-Koba
- ☐ Oban
- ☐ Odzala
- ☐ Okwangwo
- ☐ Omo
- ☐ Oueme Lower Valley
- ☐ Queen Elizabeth
- ☐ Quirimbas
- ☐ Sahamalaza – Iles Radama
- ☐ Samba Dia
- ☐ Sena Oura
- ☐ Serengeti-Ngorongoro
- ☐ Sheka
- ☐ Songor
- ☐ Taï
- ☐ The Island of Principe
- ☐ Tsimanampesotse – Nosy Ve Androka
- ☐ Vhembe
- ☐ Volcans
- ☐ Waterberg
- ☐ Waza
- ☐ Yangambi
- ☐ Yayu

Page 02

3. What do you think should be prioritised within your local context: climate change mitigation, or adaptation?

Please also state the reason for your answer.

4. Which UNESCO biosphere reserve zone should be prioritised for climate change adaptation and mitigation projects to have the biggest impact?

- ☐ Core zone
- ☐ Buffer zone
- ☐ Transition zone
- ☐ Please provide a reason for your awnser.

5. How resilient would you classify the natural environment of these UNESCO Biosphere Reserve(s) to be, to a 2°C climate increase?

Resilience is defined as "the magnitude of disturbance that can be absorbed before the system changes the variables and processes that control behaviour".

Please classify the degree of resilience: [Please choose] ⌄

6. What you think will be the impact 2°C climate increase on the different zones of the UNESCO biosphere reserve(s) you represent or have intimate knowledge of?

Please discuss in detail.

Core zone

Buffer

Transition

7. Has there been research conducted on the impact of climate change on the UNESCO biosphere reserve(s) you are associated with?

Indicate the biosphere reserve that you know has done this research.

- ☐ Aïr et Ténéré
- ☐ Amboseli
- ☐ Badiar
- ☐ Basse-Lobaye
- ☐ Belo-sur-Mer Kirindy-Mite
- ☐ Benoué
- ☐ Bia
- ☐ Black River Gorges-Bel Ombre
- ☐ Boloma-Bijagós
- ☐ Boucle du Baoulé
- ☐ Cape West Coast
- ☐ Cape Winelands
- ☐ Chimanimani
- ☐ Comoé
- ☐ Complex W-Arly-Penjari (WAP) Transboundary
- ☐ Complexe Oti-Kéran / Oti-Mandouri
- ☐ Delta du Fleuve Sénégal Transboundary
- ☐ Delta du Saloum
- ☐ Dimonika
- ☐ Dja
- ☐ Doumba-Rey
- ☐ East Usambara
- ☐ Ferlo
- ☐ Fogo
- ☐ Gadabedji
- ☐ Garden Route
- ☐ Gishwati-Mukura Landscape
- ☐ Gombe Masito Ugalla
- ☐ Gouritz Cluster
- ☐ Hadejia Nguru Bade
- ☐ Haut Niger
- ☐ Jozani – Chwaka Bay
- ☐ Kafa
- ☐ Kafue Flats
- ☐ Kiunga
- ☐ Kogelberg
- ☐ Kruger to Canyons
- ☐ Lake Bosomtwe
- ☐ Lake Chilwa Wetland
- ☐ Lake Manyara
- ☐ Lake Tana
- ☐ Littoral de Toliara

- [ ] Lubombo
- [ ] Luki
- [ ] Magaliesberg
- [ ] Maio
- [ ] Majang Forest
- [ ] Malindi Watamu Arabuko Sokoke
- [ ] Mananara Nord
- [ ] Mare aux hippopotames
- [ ] Marico
- [ ] Massif du Ziama
- [ ] Matšeng
- [ ] Middle Zambezi
- [ ] Mono Transboundary
- [ ] Monts Nimba
- [ ] Mount Elgon
- [ ] Mount Kenya – Lewa
- [ ] Mount Kulal
- [ ] Mount Mulanje
- [ ] Mwali
- [ ] Niokolo-Koba
- [ ] Oban
- [ ] Odzala
- [ ] Okwangwo
- [ ] Omo
- [ ] Oueme Lower Valley
- [ ] Queen Elizabeth
- [ ] Quirimbas
- [ ] Sahamalaza – Iles Radama
- [ ] Samba Dia
- [ ] Sena Oura
- [ ] Serengeti-Ngorongoro
- [ ] Sheka
- [ ] Songor
- [ ] Taï
- [ ] The Island of Principe
- [ ] Tsimanampesotse – Nosy Ve Androka
- [ ] Vhembe
- [ ] Volcans
- [ ] Waterberg
- [ ] Waza
- [ ] Yangambi
- [ ] Yayu

8. As the climate changes, fauna and flora must migrate across the landscape to areas where the microclimate is more suitable for their survival. The more connected a landscape is, the easier it is for these species to migrate. When a landscape is disconnected because of fences, roads or built-up areas, it prevents species from migrating. This is referred to as landscape connectivity.

How would you classify the landscape connectivity of the UNESCO biosphere reserve(s) you represent?

[Please choose] ⌄

9. Please indicate to what extent you agree or disagree with the folllowing statement.

The different zones of the UNESCO biosphere reserve(s) are functioning as mandated by the technical guidelines of the UNESCO MAB Programme.

- ☐ Strongly agree
- ☐ Agree
- ☐ Neutral
- ☐ Disagree
- ☐ Strongly disagree
- ☐ Please provide a reason for your awnser.

10. Please indicate to what extent you agree or disagree with the following statement.

The UNESCO biosphere reserve(s) addresses environmental realities, and local needs while considering the governing body's capacity for long-term implementation, in their climate change projects.

- ☐ Strongly agree
- ☐ Agree
- ☐ Neutral
- ☐ Disagree
- ☐ Strongly disagree
- ☐ Further comments:

Page 03

11. At which UNESCO MAB level should climate change adaptation or mitigation initiatives originate for the project/initiative to be sustainable?

Please indicate:

- ☐ Local
- ☐ National
- ☐ International
- ☐ Other

12. Which stakeholder(s) should be responsible for climate change mitigation and adaptation projects, for it to be sustainable?

Please indicate:

- ☐ International Coordinating Council
- ☐ Thematic/regional network administration
- ☐ Thematic/regional network membership
- ☐ National committee administration
- ☐ National committee membership
- ☐ Biosphere reserve coordinator/chairperson
- ☐ Formal conservation area management
- ☐ Government organisation
- ☐ Biosphere reserve board member/director
- ☐ Closely affiliated company to the biosphere reserve
- ☐ Research
- ☐ Consultation/advisory
- ☐ Other

13. Which stakeholders must initiate and/or conduct climate change research for the results to be actively implemented in the UNESCO biosphere reserve(s)?

- ☐ International Coordinating Council
- ☐ Thematic/regional network administration
- ☐ Thematic/regional network membership
- ☐ National committee administration
- ☐ National committee membership
- ☐ Biosphere reserve coordinator/chairperson
- ☐ Formal conservation area management
- ☐ Government organisation
- ☐ Biosphere reserve board member/director
- ☐ Closely affiliated company to the biosphere reserve
- ☐ Research
- ☐ Consultation/advisory
- ☐ Other

14. Considering recent global events (e.g., Covid pandemic) and the expected increase of such events with climate change, how resilient are the UNESCO biosphere reserve's climate change projects?

Resilience is defined as "the magnitude of disturbance that can be absorbed before the system changes the variables and processes that control behaviour".

Please classify the degree of resilience: [Please choose]

15. Have the climate change projects adopted since 2000, thus far had any marked impact on mitigating or adapting the UNESCO biosphere reserve(s) to climate change?

Please provide details.

16. On average, what is the duration of a climate change mitigation or adaptation project in your biosphere reserve?

Please indicate the average duration of projects.

17. What is generally the intended area size of climate change adaptation or mitigation projects in your UNESCO biosphere reserve(s)?

Please indicate either as a percentage or in km².

%

Km²

18. Can you provide an estimate of the average annual cost per climate change project?

Please indicate clearly your currency.

19. Our results indicate that 49% of all climate change projects' funding originates internationally, 20% locally, 12% nationally/federally, and 11% from the state or province. What are your thoughts on this?

Page 04

20. Were you involved with the creation of any of the international agreements containing climate change goals?

21. Did you attend the Sharm el-Sheikh conference on climate change in November 2022?

[Please choose]

22. Do you foresee any changes that the outcome of the Sharm el-Sheikh conference on climate change might bring about in the UNESCO biosphere reserves you know?

Please elaborate on your answer.

23. Did you attend the 15th Conference of the Parties (COP15) to the Convention on Biological Diversity?

[Please choose]

24. At the 15th Conference of the Parties (COP15) to the Convention on Biological Diversity, it was decided to try to protect and conserve at least 30% of the biodiversity of land and oceans by 2030.

What are your thoughts on this target?

25. What do you foresee the impact of this 30x30 target will be on the UNESCO biosphere reserves you are familiar with?

26. Do you think the UNESCO biosphere reserve status and its global network have a substantial impact on climate change adaptation and mitigation?

Please provide details.

27. How would you classify the relationship between the individual biosphere reserve(s), and the national, regional, and international MAB networks?

Please classify the relationships where a strong relationship = successful mutual contributions and communications:

	Strong	Strong-medium	Medium	Medium-weak	Weak
International	▢	▢	▢	▢	▢
National	▢	▢	▢	▢	▢
Regional	▢	▢	▢	▢	▢

28. Considering the global pandemic, and the expected increase of such events because of climate change, how resilient would you say the local UNESCO biosphere reserves' institutional structures are?

Resilience is defined as "the magnitude of disturbance that can be absorbed before the system changes the variables and processes that control behaviour".

[Please choose] ⌄

29. Based on our findings, organisations independent of the biosphere reserve are very to moderately active in implementing climate change projects in all zones of the biosphere reserve. What do you think the role of these organisations will be in the future?

Please provide details.

Page 05

30. How resilient would you say are local livelihoods dependent on the environment to the effects of climate change within the UNESCO biosphere reserve(s)?

Resilience is defined as "the magnitude of disturbance that can be absorbed before the system changes the variables and processes that control behaviour".

[Please choose] ⌄

31. In general, how strong are the relationships between biosphere reserves and local populations, businesses, and institutions (e.g. schools and churches) in your country?

Where strong = successful mutual contributions and communications.

Please classify the relationship strength: [Please choose] ⌄

32. Please indicate to what extent you agree or disagree with the folllowing statement.

Local populations, businesses, and institutions (e.g. schools and churches) are aware of the impact that a temperature rise will have on them.

[Please choose] ⌄

33. Please indicate to what extent you agree or disagree with the folllowing statement.

Companies are aware of the impact that a rise in temperature will have on them.

[Please choose] ⌄

34. The following suggestions were provided by respondents from the first round of questionnaires.

In each category, what are the most important suggestions on how the UNESCO biosphere reserve network can contribute to limiting global temperature increase?

Governance

- ☐ Building capacity of institutions in climate policy review and analysis.
- ☐ Use of a network of experts in organizing country-level training/seminars around climate change topics: global temperature increase to below 1.5º C by 2050.
- ☐ Capacity building of biosphere reserves to maintain their role as learning centres for sustainable development by perfectly reconciling protection and development.
- ☐ Capacity building for biosphere reserve managers.
- ☐ Closer collaboration at all levels.
- ☐ More direct involvement and intervention from UNESCO into the 25-50 hotspot biosphere reserves and showcase this in a worldwide awareness project. UNESCO does not do enough.
- ☐ Strengthening the technical capacities of managers.
- ☐ Orientation of national decision-makers on the challenges of global warming.
- ☐ The network could be more active in each country to promote meetings, workshops, and actions to elevate this issue.

Resource system

- ☐ Strengthening the monitoring and conservation of core areas of biosphere reserves.
- ☐ Improving conservation of the buffer zone and transition of the reserve.
- ☐ Enhancing the conservation of biodiversity including forest resources.
- ☐ Strengthening the supervision of actions carried out in the buffer zones.
- ☐ Developing action in favour of the populations in zones bordering the protected areas.
- ☐ Integration of the climate change issue into feasible development projects in the transition zone.
- ☐ Designating more biosphere reserves and engaging in forest restoration of degraded lands, especially in the transition zones of the biosphere reserves.
- ☐ Contributing to the reforestation and enrichment of agricultural areas with forest species of interest (through agroforestry methods) in the transition zone.

Resource Units

- ☐ Financially support the biosphere reserve to implement projects aimed at mitigating climate change.
- ☐ Become more involved with the acquisition of funding for biosphere reserves that are reaching their goals.
- ☐ Enabling the conceptualization of bankable projects.
- ☐ Lobbying for the effective payment of carbon credits.

Actors

- ☐ By promoting the exchange of experience between biosphere reserves to share success stories on the issue of climate change on a local, regional, and international level.
- ☐ Raise awareness of biosphere reserves among the public and decision-makers.
- ☐ Raise people's awareness of the challenges and benefits of working on resilience to climate change.
- ☐ Involve local partners in the restoration of degraded areas.
- ☐ Increase the participation of local communities living near biosphere reserves in the management and conservation of natural resources and encourage them to reduce their CO2 emissions and promote sustainable agriculture.
- ☐ Intensify and support experience-sharing meetings at local, regional, and international levels.
- ☐ The government should also provide resources to support the initiatives of Indigenous people.
- ☐ Promoting the principles of prevention, precaution, and adaptation.
- ☐ Support the development of project finances and the search for partners.
- ☐ There should be a central information portal on climate change so all the efforts by individual biosphere reserves can be coordinated and shared.

Interactions

- ☐ Evaluation of the amount of CO2 sequestered by tropical forest biosphere reserves to proceed with financial compensation. This financial windfall could be used for activities compatible with the conservation of natural resources for the benefit of local populations or to finance activities to restore degraded ecosystems.
- ☐ Education should be enhanced particularly for industries emitting large quantities of CO2 (to pursue mitigating factors and an increase in the climate change adaptation projects) and at a local level to assist communities to cope with the effects of climate change.
- ☐ There should be projects that focus on quantifying the carbon footprint of the biosphere reserves, followed by projects that aim to reduce the carbon footprint, or promote activities that lead to a lower carbon footprint.
- ☐ When the MAB Secretariat or other donors make funds available to MAB National Committees for restoration, monitoring and training activities, people will be more motivated to preserve the resources.

Outcomes

- ☐ Reducing population growth should be a priority while improving livelihoods by increasing the quality of education and raising awareness of climate change issues.
- ☐ The emission of carbon dioxide into the environment must be reduced and avoided.
- ☐ Biosphere reserves should serve as field laboratories for testing practical solutions to forest ecosystem management problems.
- ☐ UNESCO has major assets that could be released to help plan activities to reconcile conservation and sustainable development.
- ☐ Create opportunities to promote research on climate change in the biosphere reserves.
- ☐ Develop a green economy in biosphere reserves by creating alternative livelihood projects.
- ☐ Create environmental education clubs and introduce environmental education in primary schools.
- ☐ Implement biodiversity-related programmes and climate change mitigation projects.
- ☐ Enforce fines on polluters.
- ☐ Empower youth organizations.
- ☐ Support renewable energy and reduce coal energy.
- ☐ More intensive food forest approach to smallholder farming.
- ☐ Prioritise secure large-scale tree planting.
- ☐ Other

35. In which climate change scenario do your organisation's efforts fit?

[Please choose]

36. Which scenario do you think is most likely to happen?

[Please choose]

Last Page

Thank you for completing this questionnaire!

We would like to thank you very much for helping us.

Your answers were transmitted, you may close the browser window or tab now.

M.A. Nina Botha, Julius-Maximilians-Universität Würzburg – 2023

Appendix C

Cronbach's Alpha questions

Questionnaire 1:

The following questions are included in the first questionnaire to determine its internal reliability. They represent 43% of all the questions asked in the questionnaire. The remaining questions do not meet the criteria that would allow a Cronbach Alpha test and are excluded from this list:

1. Are these (climate change projects funded on an international, national/ federal, state/provincial, or local level?
2. In which biosphere reserve zone(s)are climate change adaptation and mitigation projects being implemented?
3. Is the biosphere reserve directly involved in any of these projects?
4. To which extend do you agree:
   4.1. Climate change is an important issue in the management plan of the biosphere reserve.
   4.2. The biosphere reserve administration (and/or the government entity in charge) has formulated a specific climate change strategy/action plan for the region.
   4.3. The biosphere reserve has positioned itself specifically concerning climate change – it is a "hotspot" and/or climate change is its main focus.
   4.4. The biosphere reserve has a specific budget for climate change actions.
5. Were the Dresden Declaration, Aichi Biodiversity targets, Paris Agreement and/or Glasgow Pact contributing to the decision to pursue these projects?
6. How active is the organisation(s) in the biosphere reserves' core zone (independent the BR) in implementing climate change mitigation and adaptation projects?
7. How active is the organisation(s) in the biosphere reserves' buffer zone (independent the BR) in implementing climate change mitigation and adaptation projects?

8. How active is the organisation(s) in the biosphere reserves' transition zone (independent the BR) in implementing climate change mitigation and adaptation projects?
9. Which climate change scenario do you think is most likely to happen?
10. In which climate change scenario do your organisation's efforts fit?

Questionnaire 2:

The following questions are used to determine the Cronbach Alpha score of the second questionnaire. These represent 44% of the total questionnaire questions and are:

1. How resilient would you classify the natural environment of these UNESCO Biosphere Reserves to be to a 2°C temperature increase?
2. How would you classify the landscape connectivity of the UNESCO Biosphere Reserve(s) that you represent?
3. Please indicate to what extent you agree with the following statement:
    3.1. The different zones of the UNESCO Biosphere Reserve(s) are functioning as mandated by the technical guidelines of the UNESCO MAB Programme.
    3.2. The UNESCO Biosphere Reserve(s) addresses environmental realities, and local needs while considering the governing body's capacity for long term implementation, in their climate change project.
4. At which UNESCO MAB level should climate change adaptation and mitigation initiatives originate for the project/initiative to be sustainable?
5. Considering the recent global events (e.g., covid pandemic) and the expected increase of such events with climate change, how resilient are the UNESCO Biosphere Reserve's climate change projects?
6. How would you classify the relationship between the individual biosphere reserve(s) and the national, regional, and international, MAB networks?
7. Considering the global pandemic, and the expected increase of such events because of climate change, how resilient would you say the local UNESCO Biosphere Reserves' institutional structures are?
8. How resilient would you say are the local livelihoods dependent on the environment to the effects of climate change within the UNESCO Biosphere Reserve(s)?

9. In general, how strong are the relationships between biosphere reserves and local populations, businesses, and institutions (e.g., schools and churches) in your country?
10. Please indicate to what extent you agree or disagree with the following statement:
    10.1. Local populations, businesses, and institutions (e.g., schools and churches) are aware of the impact that a temperature rise will have on them.
    10.2. Companies are aware of the impact that a temperature rise will have on them.
11. Which climate change scenario do you think is most likely to happen?
12. In which climate change scenario do your organisation's efforts fit?

Appendix D

Results data tables

Table 1 UNESCO AfriMAB respondents' level of involvement by country.

Country (n=63)	International	National	Local	Other	Total
Benin	1	3	1	1	3
Botswana	0	1	1	0	1
Burkina Faso	5	6	5	1	6
Cameroon	0	2	4	0	4
Republic of Congo	0	2	2	0	2
Côte d'Ivoire	4	8	7	1	9
Democratic Republic of Congo	1	3	2	0	3
Eswatini	0	2	3	0	3
Ghana	0	1	1	0	1
Guinea - Bissau	1	2	1	0	2
Madagascar	1	0	1	0	2
Malawi	1	1	1	1	1
Mali	1	2	2	0	2
Mozambique	0	0	2	0	2
Niger	1	1	0	0	1
Nigeria	1	1	1	0	1
Rwanda	3	5	1	0	5
South Africa	3	6	12	1	13
Tanzania	0	0	1	0	1
Togo	0	1	1	0	1
Total	23	47	49	5	63

Source: Own compilation

Table 2 First questionnaire on the impact of the UNESCO status and its global network on climate change adaptation and mitigation.

Response	Frequency	Percentage
Yes	32	73%
Potentially	8	18%
No	3	7%
I do not know	1	2%
Total (n=44)	44	100%

Source: Own compilation

Table 3 Second questionnaire on the impact of the UNESCO status and its global network on climate change adaptation and mitigation.

Response	Frequency	Percentage
Yes	15	68%
Potentially	3	14%
No	3	14%
I do not know	1	5%
Total (n=22)	22	100%

Source: Own compilation

Table 4 Number of UNESCO MAB thematic zones participated in as a country.

Country (n=20)	Drylands	Mangroves	Marine, coastal and Island Areas	Mountains	Savanna	Tropical forest	Wetlands	No. of thematic zones per country
Benin		x	x		x			3
Botswana							x	1
Burkina Faso	x				x		x	3
Cameroon					x	x		2
Congo						x		1
Côte d’Ivoire	x	x	x	x	x	x	x	7
DRC		x	x	x		x	x	5
Eswatini				x	x		x	3
Ghana		x	x				x	3
Guinea - Bissau		x	x		x	x	x	5
Madagascar			x			x	x	3
Malawi				x				1
Mali					x			1
Mozambique	x	x	x	x	x	x	x	7
Niger	x				x		x	2
Nigeria		x			x	x		3
Rwanda				x		x	x	3
South Africa	x		x	x	x	x	x	6
Tanzania		x	x					2
Togo					x			1
No. of countries	5	8	9	7	12	10	12	

Source: Own compilation

Table 5 The impact of a 2°C temperature increase on the different zones of the biosphere reserve(s) within each country (Benin to Guinea-Bissau).

	(n=14)	Benin (n=1)	Burkina Faso (n=4)	Cameroon (n=2)	DRC (n=1)	Côte d'Ivoire (n=3)	Eswatini (n=1)	Guinea - Bissau (n=1)
Biosphere reserve zones	Core	Flooding of wildlife habitats leads to migration and exposure to poaching. Loss of species diversity.	Reduced water potential and diversity plus well-being of flora and fauna.	Decrease in the flow of water bodies (ponds, streams), drying up of marshy areas, leading to the destruction of certain species of fauna and flora.	-	Loss of habitats and biodiversity due to agricultural encroachment. Reduction or increase in river flows, drying up of pools, flooding, landslides, and the disappearance of species.	Increased threat to endangered species and ecosystems. Enhanced threat from the populace with the increase in poaching and illegal harvesting of protected resources.	Coastal erosion is affecting the plant and animal species of the coastline due to the physical and chemical processes of coastal dune destruction.
	Buffer	An increase in food and water crises & loss of biodiversity due to the limited capacity and speed of adaptation of fauna and flora.	Extinction of resources exploitable by local populations. Reduction in water potential, decrease in the diversity and well-being of fauna and flora species, and population pressure.	Increase in human pressure on bioresources, advancing agricultural front, and regular human-wildlife conflicts.	-	Development of inappropriate production systems. Increased anthropic pressure, and increased conflicts with riverside populations.	Increased poverty and dependence on biodiversity as ecosystem-dependent livelihoods are negatively impacted.	-

Biosphere reserve zones	(n=14)	Benin (n=1)	Burkina Faso (n=4)	Cameroon (n=2)	DRC (n=1)	Côte d'Ivoire (n=3)	Eswatini (n=1)	Guinea - Bissau (n=1)
	Tran-si-tion	Degradation of certain dwellings may cause the migration of populations from their original environment. Loss of cultural and cultic values.	Reduction in water potential, decrease in the diversity and well-being of fauna and flora species, and population pressure on the resources of the transition zone. Along with soil impoverishment and a decrease in agro-salvo-pastoral production.	Increasing conflicts between different stakeholders over access to resource areas. Loss of wildlife and its habitat, together with a decline in the agricultural production of the people.	Savannahisation and the resurgence of fires in these degraded areas, an increase in the magnitude of storms causing deadly floods, and a decrease in agricultural yields.	Abandonment of non-fertile agricultural land. Reduction in agricultural production, degradation of land, displacement of populations, emergence of diseases, reduction of water resources.	Increased poverty and dependence on biodiversity as ecosystem-dependent livelihoods are negatively impacted.	-
	Climatic pre-dic-tion	Increase droughts & extreme events	Increased drought	Increased droughts	Increase droughts & extreme events	Increased droughts and extreme events	-	Coastal erosion

Source: Own compilation

Table 6 The impact of a 2°C temperature increase on the different zones of the biosphere reserve(s) within each country (Malawi to Tanzania).

Biosphere reserve zones (n=14)	Malawi (n=1)	Mali (n=1)	Mozambique (n=1)	Niger (n=1)	Rwanda (n=1)	South Africa (n=6)	Tanzania (n=1)
Core	Increased wildfire damage and flash floods from intense storms.	The acceleration of the drying up of water points, and the increase in anthropic pressure on the main areas (core areas).	Loss of Biodiversity - alteration of habitats and ecosystems.	This would be a continuous and accelerated degradation of natural resources and biodiversity or even the disappearance of certain fauna and flora species. This is currently the case here species are migrating from the north to the south, changing their habitat, and ecological niche.	Shifts in plant phenology patterns and ranges, shifts in animal ranges/distributions.	All biodiversity is currently at its temperature limit and will cause extinctions. In our biosphere, we experience heat waves. In previous heatwaves some animals in the core zone (nature and game reserves) died as a result. I think there might be an increase in droughts, wildfires, and floods' frequency and intensity.	Most of the biodiversity is scattered because of the increase in climate.
Buffer	Increased wildfire damage and flash floods from intense storms	Loss of biodiversity due to pressure from communities bordering the reserve.	Loss of Biodiversity - alteration of habitats and ecosystems.	The impact of climate change variability means that this buffer zone no longer functions as intended and negatively impacts the core area. This degradation is directly felt in the core area that this zone is supposed to protect.	Disturbances to the land cover and land use (erosion risk increases).	Heatwaves, droughts, and floods will have a negative impact on local agriculture, livestock, tourism and infrastructure. Also, natural systems, watercourses, and vegetation composition is likely to be affected.	Decrease in biodiversity and forest cover.

	(n=14)	Malawi (n=1)	Mali (n=1)	Mozambique (n=1)	Niger (n=1)	Rwanda (n=1)	South Africa (n=6)	Tanzania (n=1)
Biosphere reserve zones	Transition	Household and crop damage from floods. Crop vulnerability from unusual seasonal rains.	Loss of biodiversity due to increase in agricultural areas, drying up of certain water points, and deforestation.	Alteration of habitats and ecosystems, loss of biological diversity, increasing poverty and increasing scarcity of natural resources.	At this level the degradation is complete! the phenomenon is at its most extreme.	Shifts in flora and fauna, and increased disturbances due to unpredictable and heavy rain events.	Agriculture, human settlements, and other industry may be impacted by extreme weather events, and groundwater shortage because of drought.	Increasing conflict between locals because of low agricultural production in the area.
	Climatic prediction:	Extreme events	Increase in drought	-	Changing microclimates	Changing microclimate and increasing extreme events	Increased droughts and extreme events	Changing microclimate

Source: Own compilation

Table 7 First questionnaire, temperature increase scenarios according to respondents.

	Responses	
Climate change scenarios	N	Percentage
Scenario 1, Business as usual: temperature increase of more than 2°C by 2050.	12	24%
Scenario 2, Middling target: temperature increase remains just below 2°C by 2050.	29	58%
Scenario 3, Ideal target: temperature plateau between 1° C and 1.5°C by 2050.	9	18%
Total (n=50)	50	100%

Source: Own compilation

Table 8 Second questionnaire, temperature increase scenarios according to respondents.

	Responses	
Climate change scenarios	N	Percentage
Scenario 1, Business as usual: temperature increase of more than 2°C by 2050.	4	16%
Scenario 2, Middling target: temperature increase remains just below 2°C by 2050.	16	64%
Scenario 3, Ideal target: temperature plateau between 1° C and 1.5°C by 2050.	5	20%
Total (n=25)	25	100%

Source: Own compilation

Table 9 Countries represented in the first questionnaire.

	Countries represented in the questionnaire																					
UNESCO AfriMAB Biosphere Reserves represented	Biosphere reserves (n=33)	Benin	Botswana	Burkina Faso	Cameroon	Congo	Côte d' Ivoire	DRC	Eswatini	Ghana	Guinea - Bissau	Madagascar	Malawi	Mali	Mozambique	Niger	Nigeria	Rwanda	South Africa	Tanzania	Togo	Total per reserve
	Aïr et Ténéré															1						1
	Benoué				1																	1
	Boloma-Bijagós										1											1
	Boucle du Baoulé													2								2
	Cape West Coast																		2			2
	Cape Winelands																		1			1
	Comoé						5															5
	WAP	1		4																		7
	Dja				3																	3
	Gadabedji															1						1
	Gishwati-Mukura Landscape																	3				3
	Gouritz Cluster																		2			2
	Jozani - Chwaka Bay																			1		1
	Kogelberg																		1			1
	Kruger to Canyons																		2			2
	Lubombo								3													3
	Luki							3														3
	Magaliesberg																		2			2
	Mare aux hippo-potames			3																		3

	Countries represented in the questionnaire																				
	Biosphere reserves (n=33)	Benin	Botswana	Burkina Faso	Cameroon	Congo	Côte d' Ivoire	DRC	Eswatini	Ghana	Guinea - Bissau	Madagascar	Malawi	Mali	Mozambique	Niger	Nigeria	Rwanda	South Africa	Tanzania	Togo	Total per reserve
UNESCO AfriMAB Biosphere Reserves represented	Marico																		1			1
	Mono Trans-boundary	1																			1	2
	Mount Mulanje												1									1
	Odzala					2																2
	Omo																1					1
	Oueme Lower Va-lley	1																				1
	Quirimbas														2							2
	Sahamalaza - Iles Radama											1										1
	Songor									1												1
	Taï						5															5
	Tsimanam-pesotse - Nosy Ve Androka											1										1
	Vhembe																		3			3
	Volcans																	3				3
	Yangambi							2														2
	Makgadikgadi Wetlands		1																			1
	Total respon-dents	3	1	6	4	2	9	3	3	1	2	2	1	2	2	1	1	5	13	1	1	

Source: Own compilation

Table 10 Countries represented in the second questionnaire.

	Countries represented in the questionnaire																	
	Biosphere reserve (n=35)	Benin	Burkina Faso	Cameroon	Côte d' Ivoire	DRC	Eswatini	Ethiopia	Guinea - Bissau	Malawi	Mali	Mozambique	Niger	Nigeria	Rwanda	South Africa	Tanzania	Total
UNESCO AfriMAB Biosphere Reserves	Aïr et Ténéré												1					1
	Benoué			1														1
	Boloma-Bijagós								1									1
	Boucle du Baoulé										1							1
	Cape West Coast															1		1
	Cape Winelands															1		1
	Comoé				2													2
	WAP	1	3										1					5
	Dja			2														2
	Doumba-Rey			1														1
	East Usambara																1	1
	Gadabedji												1					1
	Gishwati-Mu-kura Landscape														1			1
	Gouritz Cluster															1		1
	Jozani - Chwaka Bay																1	1

	Countries represented in the questionnaire																	
	Biosphere reserve (n=35)	Benin	Burkina Faso	Cameroon	Côte d' Ivoire	DRC	Eswatini	Ethiopia	Guinea - Bissau	Malawi	Mali	Mozambique	Niger	Nigeria	Rwanda	South Africa	Tanzania	Total
UNESCO AfriMAB Biosphere Reserves	Kafa							1										1
	Kogelberg															1		1
	Kruger to Canyons															2		2
	Lake Manyara																1	1
	Lubombo						1											1
	Magaliesberg															2		2
	Majang Forest							1										1
	Mare aux hip-popotames		2															2
	Mono Trans-boundary	1																1
	Mount Mulanje									1								1
	Omo													1				1
	Quirimbas											2						2
	Serengeti-Ngorongoro																1	1
	Sheka							1										1
	Taï				5													5
	Vhembe															2		2

	Countries represented in the questionnaire																	
UNESCO AfriMAB Biosphere Reserves	Biosphere reserve (n=35)	Benin	Burkina Faso	Cameroon	Côte d' Ivoire	DRC	Eswatini	Ethiopia	Guinea - Bissau	Malawi	Mali	Mozambique	Niger	Nigeria	Rwanda	South Africa	Tanzania	Total
	Volcans														2			2
	Waterberg															2		2
	Waza			1														1
	Yangambi					1												1
	Total	2	4	2	5	1	1	1	1	1	1	2	1	1	2	7	1	33

Source: Own compilation

Table 11 AfriMAB Biosphere Reserves represented with climate change related projects.

	Countries represented in the questionnaire																			
Biosphere reserves (n=30)	Benin	Botswana	Burkina Faso	Cameroon	Congo	Côte d' Ivoire	DRC	Eswatini	Ghana	Guinea - Bissau	Madagascar	Malawi	Mali	Mozambique	Niger	Nigeria	Rwanda	South Africa	Tanzania	Togo
Aïr and Ténéré															√					
Benoué				x																
Boloma-Bi-jagós										x										
Boucle du Baoulé													√							
Cape West Coast																		x		
Cape Winelands																		√		
Comoé						√														
WAP	√		√												√					
Dimonika					√															
Dja				√																
Gadabedji															√					
Gishwati-Mukura Landscape																	√			
Gouritz Cluster																		√		
Kogelberg																		x		
Kruger to Canyons																		√		
Lubombo								√												
Luki							√													
Magalies-berg																		√		

	Countries represented in the questionnaire																			
Biosphere reserves (n=30)	Benin	Botswana	Burkina Faso	Cameroon	Congo	Côte d' Ivoire	DRC	Eswatini	Ghana	Guinea - Bissau	Madagascar	Malawi	Mali	Mozambique	Niger	Nigeria	Rwanda	South Africa	Tanzania	Togo
Mare aux hippo-potames			√																	
Marico																		√		
Mono Trans-boundary	√																			√
Mount Mulanje												√								
Odzala					√															
Omo																√				
Oueme Lo-wer Valley	√																			
Quirimbas														√						
Sahamalaza - Iles Radama											√									
Songor									√											
Serengeti-Ngorongoro																			√	
Taï						√														
Tsimanam-pesotse - Nosy Ve Androka											√									
Vhembe																		√		
Volcans																	√			
Waterberg																		√		
Yangambi							√													
Makgadik-gadi Wetlands		√																		
Index: Yes, there is climate-related projects √ No there is not x																				

Source: Own compilation

Table 12 AfriMAB Biosphere Reserves represented and average climate change projects since 2000.

	Biosphere reserves (n=27)	Benin	Burkina Faso	Cameroon	Congo	Côte d' Ivoire	DRC	Eswatini	Ghana	Madagascar	Mali	Mozambique	Niger	Rwanda	South Africa	Tanzania	Togo
UNESCO AfriMAB Biosphere Reserves	Boucle du Baoulé										2						
	Cape West Coast														0		
	Cape Wine-lands														4		
	Comoé					3											
	WAP	3	1										1				
	Dja			0													
	Gadabedji												5				
	Gishwati-Mukura Landscape													2			
	Gouritz Cluster														4		
	Jozani - Chwaka Bay															4	
	Kogelberg														0		
	Kruger to Canyons														2		
	Lubombo							5									
	Luki						4										
	Magalies-berg														2		
	Mare aux hippopota-mes		2														
	Mono Trans-boundary	3															3

	Biosphere reserves (n=27)	Benin	Burkina Faso	Cameroon	Congo	Côte d' Ivoire	DRC	Eswatini	Ghana	Madagascar	Mali	Mozambique	Niger	Rwanda	South Africa	Tanzania	Togo
UNESCO AfriMAB Biosphere Reserves	Odzala				1												
	Oueme Lower Valley	5															
	Quirimbas								1			2					
	Songor																
	Sahamalaza - Iles Radama									1							
	Taï					2											
	Tsimanampesotse - Nosy Ve Androka									5							
	Vhembe														2		
	Volcans													3			
	Yangambi						4										
	Total projects	11	3	0	1	5	8	5	1	6	2	2	6	5	14	4	3

Source: Own compilation

Table 13 Zones where most climate change projects are being implemented.

Zones where most climate change projects are being implemented				
UNESCO AfriMAB Biosphere Reserves	Biosphere reserves (n=30)	Core	Buffer	Transi-tion
	Aïr et Ténéré	x	x	x
	Boucle du Baoulé			x
	Cape West Coast	x	x	x
	Cape Winelands	x	x	x
	Comoé	x	x	x
	WAP	x	x	x
	Dja	x		x
	Gadabedji	x	x	x
	Gishwati-Mukura Landscape	x	x	x
	Jozani - Chwaka Bay	x	x	
	Gouritz Cluster	x	x	x
	Kogelberg			x
	Kruger to Canyons	x	x	x
	Lubombo	x	x	x
	Luki	x	x	x
	Magaliesberg	x	x	x
	Mare aux hippopotames	x	x	x
	Marico	x	x	x
	Mono Transboundary		x	x
	Mount Mulanje			x
	Odzala		x	
	Omo	x	x	x

UNESCO AfriMAB Biosphere Reserves	Zones where most climate change projects are being implemented			
	Biosphere reserves (n=30)	Core	Buffer	Transi-tion
	Quirimbas	x	x	x
	Sahamalaza - Iles Radama	x		
	Songor	x	x	
	Taï	x	x	x
	Tsimanampesotse - Nosy Ve Androka	x	x	
	Vhembe	x	x	x
	Volcans		x	x
	Yangambi		x	x
	Total	23	25	25

Source: Own compilation

Table 14 Biosphere reserves that have conducted climate change impact studies.

	Representative countries in the UNESCO AfriMAB network														
Biosphere reserves that have conducted climate change impact studies	Biosphere reserves (n=20)	Benin	Burkina Faso	Cameroon	DRC	Côte d' Ivoire	Ethiopia	Guinea - Bissau	Malawi	Mali	Niger	Nigeria	Rwanda	South Africa	Tanzania
	Benoué			x											
	Boloma-Bijagós						x	x							
	Boucle du Baoulé									x					
	Comoé					x									
	WAP	x	x								x				
	Jozani - Chwaka Bay														x
	Kruger to Canyons													x	
	Mount Mulanje								x						
	Omo											x			
	Taï					x									
	Vhembe													x	
	Volcans												x		
	Waza			x											
	Yangambi				x										
	Yayu						x								
	Total	1	1	2	1	2	1	1	1	1	1	1	1	2	1

Source: Own compilation

Table 15 AfriMAB Biosphere Reserves represented.

	Carbon dioxide (CO_2) assessment conducted		
UNESCO AfriMAB Biosphere Reserves represented	Biosphere reserves (n=30)	Yes	No
	Aïr et Ténéré		x
	Boucle du Baoulé		x
	Cape West Coast		x
	Cape Winelands		x
	Comoé	x	
	WAP	x	
	Dja	x	
	Gadabedji		x
	Gishwati-Mukura Landscape		x
	Gouritz Cluster	x	
	Jozani - Chwaka Bay		x
	Kogelberg		x
	Kruger to Canyons		x
	Lubombo	x	
	Luki	x	
	Magaliesberg		x
	Mare aux hippopotames		x
	Marico		x
	Mono Transboundary		x
	Mount Mulanje		x
	Odzala		x
	Omo	x	
	Quirimbas		x
	Sahamalaza - Iles Radama		x
	Songor		x
	Taï	x	
	Tsimanampesotse - Nosy Ve Androka		x
	Vhembe		x
	Volcans		x
	Yangambi	x	
	Total	9	21

Source: Own compilation

Table 16 Resource system performance measures.

			Natural environment temp. increase resilience					Landscape connectivity					Zone functioning as intended by UNESCO MAB						
	Country code	Biosphere reserves (n=34)	High (5)	High-medium (4)	Medium (3)	Medium-low (2)	Low (1)	High (5)	High-medium (4)	Medium (3)	Medium-low (2)	Low (1)	Strongly agree (5)	Agree (4)	Neutral (3)	Disagree (2)	Strongly disagree (1)	Total	Percentage
UNESCO AfriMAB Biosphere Reserves represented	BF	Mare aux hippopotames		x						x				x				11	73%
	BJ, TG	Mono Transboundary				x					x			x				8	53%
	BJ, BF, NE	WAP			x					x				x				10	67%
	CD	Yangambi			x			x						x				12	80%
	CI	Comoé		x						x				x				11	73%
	CI	Taï		x							x			x				10	67%
	CM	Benoué			x					x						x		8	53%
	CM	Dja			x					x					x			9	60%
	CM	Doumba-Rey			x					x						x		8	53%
	CM	Waza			x					x						x		8	53%
	ET	Kafa		x					x						x			11	73%
	ET	Majang Forest		x					x						x			11	73%
	ET	Sheka		x					x						x			11	73%
	GW	Boloma-Bijagós		x							x				x			9	60%
	ML	Boucle du Baoulé				x				x			x					10	67%
	MW	Mount Mulanje				x					x			x				8	53%
	MZ	Quirimbas				x				x				x				9	60%

			Natural environment temp. increase resilience					Landscape connectivity					Zone functioning as intended by UNESCO MAB						
	Country code	Biosphere reserves (n=34)	High (5)	High-medium (4)	Medium (3)	Medium-low (2)	Low (1)	High (5)	High-medium (4)	Medium (3)	Medium-low (2)	Low (1)	Strongly agree (5)	Agree (4)	Neutral (3)	Disagree (2)	Strongly disagree (1)	Total	Percentage
UNESCO AfriMAB Biosphere Reserves represented	NE	Aïr et Ténéré					x							x				5	33%
	NE	Gadabedji					x							x				5	33%
	NG	Omo				x		x					x					12	80%
	RW	Gishwati-Mukura Landscape				x						x			x			6	40%
	RW	Volcans			x						x				x			8	53%
	SZ	Lubombo					x				x				x			6	40%
	TZ	East Usambara				x			x					x				10	67%
	TZ	Jozani - Chwaka Bay				x			x					x				10	67%
	TZ	Lake Manyara				x			x					x				10	67%
	TZ	Serengeti-Ngorongoro				x			x					x				10	67%
	ZA	Cape West Coast					x		x							x		7	47%
	ZA	Gouritz Cluster					x			x			x					9	60%
	ZA	Kogelberg				x				x				x				9	60%
	ZA	Kruger to Canyons				x				x				x				9	60%
	ZA	Magaliesberg				x				x					x			8	53%
	ZA	Vhembe					x			x				x				8	53%
	ZA	Waterberg		x						x				x				11	73%
		Total	0	8	7	13	6	2	8	15	6	1	3	18	9	3	0		

Source: Own compilation

Table 17 Climate change project categories.

Climate change project categories (n=29)	Aïr et Ténéré	Boucle du Baoulé	Cape West Coast	Cape Winelands	WAP	Dja	Gadabedji	Gishwati-Mukura Landscape	Gouritz Cluster	Jozani - Chwaka Bay	Kogelberg	Kruger to Canyons	Lubombo	Luki	Magaliesberg	Mare aux hippopotames	Marico	Mono Transboundary	Mount Mulanje	Odzala	Omo	Quirimbas	Sahamalaza - Iles Radama	Songor	Taï	Tsimanampesotse - Nosy Ve Androka	Vhembe	Volcans	Yangambi	Total
Forestry that mitigates climate change including reduced deforestation and reforestation	x	x			x	x	x			x	x		x	x	x	x			x		x	x		x	x		x		x	18
Adaptation to climate change in forestry	x	x			x	x	x			x				x		x		x	x	x	x	x			x		x		x	16
Mainstreaming biodiversity	x	x	x		x		x		x				x	x	x	x		x	x			x			x				x	15
Mitigation of climate change in agriculture and/or husbandry	x				x	x	x		x	x	x		x	x		x		x				x			x				x	14
Enhancing landscape connectivity through area-based conservation and/or “other effective area-based conservation measures’ that achieve conservation goals outside formally protected areas (e.g., biodiversity corridors)	x	x			x	x	x						x	x	x	x					x				x	x	x		x	14
Restoring degraded (semi-) arid ecosystems	x	x	x		x		x	x	x				x				x				x					x	x	x	x	14

Adaptation to climate change in agriculture and/or husbandry (including the revitalization of traditional farming knowledge)		x			x				x	x	x	x	x	x		x	x				x				x				x	13
Balancing conflicting interests and newly competing forms of land use	x	x			x		x		x			x	x	x	x			x			x				x				x	13
Enhancing biodiversity conservation in transformed ecosystems	x				x	x	x		x	x				x		x		x	x		x				x				x	13
Mitigating the residual impact of development (e.g., mining, urban development, agricultural expansion, green agriculture initiatives, sustainable forestry, hunting, and anti-poaching programmes)	x	x			x	x	x		x				x	x	x	x			x		x								x	13
Sustainable fishing	x		x		x	x	x			x			x			x		x				x	x			x				12
Reducing pollution (all sources)		x		x	x	x			x			x	x	x	x						x								x	11
Increasing the formal area under protection	x					x	x		x				x		x		x	x							x	x				10
Adaptation to climate change in river/flood management		x		x	x				x			x	x			x		x									x			9
Improved stoves for simple solid biomass		x	x		x								x	x				x	x		x								x	9
Adaptation to specifically threatened species	x		x		x		x						x		x							x			x	x				9

Rewilding with large terrestrial mammals	x				x		x		x				x			x		x							x					8
Inland wetland restoration	x		x		x		x		x				x					x									x			8
Adaptation through improved mosaics of natural and cultivated spaces	x	x			x				x					x		x													x	7
Adaptation to increasing water scarcity/droughts/desertification	x	x		x	x				x				x			x														7
Implementing emission reduction trading mechanisms												x	x	x						x	x								x	6
Sustainable consumption patterns					x								x	x				x			x								x	6
Adaptation in the tourism industry (e.g., low-impact tourism)	x				x		x			x			x				x													6
Using fire and bush removal to combat woody plant encroachment	x			x	x		x										x								x					6
Coastal zone adaptation (sea-level rise, storm surges, hurricanes, groundwater salinity)			x							x				x										x					x	5
Eliminating subsidies harmful to biodiversity													x								x						x			3

Adaptation to new species (migratory and/or invasive)			x										x		x															3
Improving energy efficiency in buildings and cities					x																x									2
Avoiding degradation of permafrost areas	x						x																							2
Low-carbon macro-economic approaches (regional marketing and circular economic flows)												x																		1
Promotion of non-conflictual biofuels (e.g., agricultural residues or grass cuttings)													x																	1
Adaptation to shifting vegetation			x																											1
Total	19	13	9	4	23	9	17	1	15	9	4	7	24	16	10	14	6	13	7	3	15	7	2	3	13	6	8	2	17	

Source: Own compilation

Table 18 Year climate change started being pursued in AfriMAB.

Year climate change started being pursued in the UNESCO AfriMAB network																						
Country code	Biosphere reserve (n=23)	1979	1984	1999	2000	2002	2004	2007	2008	2009	2010	2012	2013	2014	2015	2016	2017	2018	2019	2020	2021	2022
BF	Mare aux hippopotames								x													
BJ, BF, NE	WAP									x												
BJ, TG	Mono Transboundary																x					
CG	Odzala					x																
CI	Taï				x																	
CL	Comoé											x										
CM	Dja		x																			
DC	Luki	x											x									
DC	Yangambi	x																				
MG	Sahamalaza - Iles Radama																	x				
ML	Boucle du Baoulé																x					
MW	Mount Mulanje							x														
MZ	Quirimbas						x															
RW	Gishwati-Mukura Landscape																			x		
RW	Volcans																			x		
SZ	Lubombo			x																		
ZA	Cape Winelands																				x	
ZA	Gouritz Cluster											x										
ZA	Kogelberg																x					
ZA	Kruger to Canyons																		x			
ZA	Magaliesberg														x							

Year climate change started being pursued in the UNESCO AfriMAB network																						
Country code	Biosphere reserve (n=23)	1979	1984	1999	2000	2002	2004	2007	2008	2009	2010	2012	2013	2014	2015	2016	2017	2018	2019	2020	2021	2022
ZA	Marico										x											
ZA	Vhembe									x												
	Total	2	1	1	1	1	1	1	1	2	1	2	1	0	1	0	3	1	1	2	1	0

Source: Own compilation

Table 19 Area covered by climate change projects.

Percentage area covered by climate change projects												
Country code	Biosphere reserves (n=23)	1-10%	11-20%	21-30%	31-40%	41-50%	51-60%	61-70%	71-80%	81-90%	91-100%	100%+
BF	Mare aux hippopotames										192	
BF, BJ, NE	WAP						53,619.90					
CM	Benoué						4,202					
CM	Dja						7,968					
CM	Doumba-Rey						1,209.60					
CM	Waza						429.552					
ML	Boucle du Baoulé				8,750							
MW	Mount Mulanje				25.68							
NE	Aïr et Ténéré				95,880							
NE	Gadabedji				7,350.85							

Percentage area covered by climate change projects												
Country code	Biosphere reserves (n=23)	1-10%	11-20%	21-30%	31-40%	41-50%	51-60%	61-70%	71-80%	81-90%	91-100%	100%+
SW	Lubombo								2,205.15			
TZ	East Usambara						503.96					
TZ	Jozani - Chwaka Bay						110.62					
TZ	Lake Manyara						1,803.15					
TZ	Serengeti-Ngorongoro						11,986.52					
ZA	Cape West Coast											10,000
ZA	Gouritz Cluster											100,000
ZA	Kogelberg	103.62										
ZA	Magaliesberg	71.57										
ZA	Vhembe								24,080.80			
* USD per year as of May 2023												

Source: Own compilation

Table 20 Duration of climate change projects.

		Duration of climate change related projects						
Country code	Biosphere reserve (n=26)	1	2	3	4	5	10 to 19	20+
BF	Mare aux hippopotames				x			
BJ, BF, NE	WAP				x			
CD	Yangambi						x	
CI	Comoé		x					
CI	Taï				x			
CM	Benoué					x		
CM	Dja						x	
CM	Doumba-Rey					x		
CM	Waza					x		
ML	Boucle du Baoulé		x					
MW	Mount Mulanje							x
NE	Aïr et Ténéré			x				
NE	Gadabedji			x				
RW	Gishwati-Mukura Landscape	x						
RW	Volcans	x						
SZ	Lubombo		x					
TZ	East Usambara				x			
TZ	Jozani - Chwaka Bay				x			
TZ	Lake Manyara				x			
TZ	Serengeti-Ngorongoro				x			
ZA	Cape West Coast	x						
ZA	Gouritz Cluster				x			
ZA	Kogelberg	x						
ZA	Kruger to Canyons				x			

		Duration of climate change related projects						
Country code	Biosphere reserve (n=26)	1	2	3	4	5	10 to 19	20+
ZA	Vhembe			x				
ZA	Waterberg							x
	Total	4	3	3	9	3	2	2

Source: Own compilation

Table 21 Average annual climate change cost estimate.

Country code	Biosphere reserves (n=16)	Average annual climate change cost estimate
BF	Mare aux hippopotames	$245.83
BJ, BF, NE	WAP	$193,888.78
CI	Comoé	$16,388.88
CI	Taï	$544,341.94
ML	Boucle du Baoulé	$163,888.77
NE	Aïr et Ténéré	$60,000.00
NE	Gadabedji	$60,000.00
TZ	East Usambara	$1,057,082.45
TZ	Jozani - Chwaka Bay	$1,057,082.45
TZ	Lake Manyara	$1,057,082.45
TZ	Serengeti-Ngorongoro	$1,057,082.45
ZA	Gouritz Cluster	$1,072,495.00
ZA	Kogelberg	$5,170.20
ZA	Kruger to Canyons	$206,808.00
ZA	Vhembe	$15,509.58
ZA	Waterberg	$15,509,577.00
* USD per year as of May 2023		

Source: Own compilation

Table 22 Estimated project duration, cost, and area.

Annual estimated cost of climate change projects USD per km²

		Percentage area covered by projects				
Country code	Biosphere reserves (n=7)	31-40%	51-60%	71-80%	91-100%	Duration of projects in years
NE	Aïr et Ténéré	\$0.63 per km²				3
ML	Boucle du Baoulé	\$18.73 per km²				2
BJ, BF, NE	WAP		\$3.62 per km²			4
NE	Gadabedji	\$8.16 per km²				3
BF	Mare aux hippopotames				\$1.28 per km²	4
ZA	Gouritz Cluster				\$10.72 per km²	4
ZA	Vhembe			\$0.64 per km²		3

Source: Own compilation

Table 23 Resource unit performance measure.

	Country code	Biosphere reserve (n=23)	The positive impact of projects since 2000			Project Resilience						
			Yes (2)	Maybe (1)	No (0)	High (5)	High-medium (4)	Medium (3)	Medium-low (2)	Low (1)	Total	Percentage
UNESCO AfriMAB Biosphere Reserves represented	BF	Mare aux hippopotames	x				x				6	86%
	BJ, BF, NE	WAP		x				x			4	57%
	CD	Yangambi			x					x	1	14%
	CI	Comoé			x			x			3	43%
	CI	Taï		x				x			4	57%
	CM	Benoué	x					x			5	71%
	CM	Dja	x					x			5	71%
	CM	Doumba-Rey	x					x			5	71%
	CM	Waza	x					x			5	71%
	ET	Kafa			x		x				4	57%
	ET	Majang Forest		x			x				5	71%
	ET	Sheka			x		x				4	57%
	ML	Boucle du Baoulé	x			x					7	100%
	MW	Mount Mulanje		x				x			4	57%
	NE	Aïr et Ténéré			x					x	1	14%
	NE	Gadabedji			x					x	1	14%
	NG	Omo	x			x					7	100%
	RW	Gishwati-Mukura Landscape		x					x		3	43%
	RW	Volcans	x					x			5	71%

	Country code	Biosphere reserve (n=23)	The positive impact of projects since 2000			Project Resilience						
			Yes (2)	Maybe (1)	No (0)	High (5)	High-medium (4)	Medium (3)	Medium-low (2)	Low (1)	Total	Percentage
UNESCO AfriMAB Biosphere Reserves represented	SZ	Lubombo		x					x		3	43%
	TZ	East Usam-bara		x			x				5	71%
	TZ	Jozani - Chwaka Bay		x			x				5	71%
	TZ	Lake Manyara		x			x				5	71%
	TZ	Serengeti-Ngorongoro		x			x				5	71%
	ZA	Cape West Coast		x					x		3	43%
	ZA	Gouritz Clus-ter			x			x			3	43%
	ZA	Kogelberg			x				x		2	29%
	ZA	Kruger to Canyons	x					x			5	71%
	ZA	Magaliesberg			x		x				4	57%
	ZA	Vhembe			x				x		2	29%
	ZA	Waterberg			x		x				4	57%
		Total	9	11	9	2	10	11	5	3	4	57%

Source: Own compilation

Table 24 Institutionalisation scores of biosphere reserves in the AfriMAB.

Biosphere reserves (n=30)	The biosphere reserve has created a budget for climate change projects				Climate change is an important issue in their management plan				Formulated climate change strategy/action plan				The biosphere reserve has positioned itself specifically concerning climate change				Institutionalisation score	
	Fully correct (3)	Mostly correct (2)	Partially correct (1)	Not correct (0)	Fully correct (3)	Mostly correct (2)	Partially correct (1)	Not correct (0)	Fully correct (3)	Mostly correct (2)	Partially correct (1)	Not correct (0)	Fully correct (3)	Mostly correct (2)	Partially correct (1)	Not correct (0)	Total (_/12)	Percentage
Aïr et Ténéré				0	3					2					1		6	50%
Boucle du Baoulé				0	3						1			2			6	50%
Cape West Coast				0	3				3						1		7	58%
Cape Winelands			1			2				2			3				8	67%
Comoé		2			3					2			3				10	83%
WAP			1			2				2				2			7	58%
Dja				0		2					1				1		4	33%
Gadabedji				0	3					2					1		6	50%

Gishwati-Mukura Landscape		2			3				3					2			10	83%
Gouritz Cluster	3				3					2			3				11	92%
Jozani - Chwaka Bay			1		3					2				2			8	67%
Kogelberg				0	3												3	25%
Kruger to Canyons		2			3					2				2			9	75%
Lubombo			1		3					2					1		7	58%
Luki			1		3					2				2			8	67%
Magaliesberg			1		3				3					2			9	75%
Mare aux hippopotames			1		3				3					2			9	75%
Marico				0	3					2					1		6	50%
Mono Transboundary			1			2					1				1		5	42%

Mount Mu-lanje			1			2						0		2			5	42%
Odzala				0		2				2						0	4	33%
Omo				0		2				2			3				7	58%
Quirimbas			1		3						1		3				8	67%
Sahamalaza - Iles Radama			1		3				3				3				10	83%
Songor			1			2				2					1		6	50%
Taï			1		3					2			3				9	75%
Tsimanam-pesotse - Nosy Ve Androka				0			1				1				1		3	25%
Vhembe			1		3					2				2			8	67%
Volcans		2			3				3					2			10	83%
Yangambi		2			3				3								8	67%
Number of votes	1	5	14	10	21	8	1	0	7	16	5	0	7	11	9	1		

Source: Own compilation

Table 25 Biosphere reserve's role in climate change projects.

Biosphere reserve's roles in climate change projects							
	Country code	Biosphere reserve (n=28)	Leading role	Partner role	Coordi-nating role	Obser-ver role	No role
UNESCO AfriMAB Biosphere Reserves represented	BF	Mare aux hippopotames	x	x	x		
	BJ, BF, NE	WAP	x	x	x	x	
	BJ, TG	Mono Transboundary		x			
	CD	Luki	x	x			
	CD	Yangambi	x	x			
	CI	Comoé	x	x	x		
	CI	Taï	x	x	x		
	CM	Dja	x	x	x		
	GH	Songor		x			
	MG	Sahamalaza - Iles Radama			x		
	MG	Tsimanampesotse - Nosy Ve Androka	x	x			
	ML	Boucle du Baoulé		x	x		
	MW	Mount Mulanje	x				
	MZ	Quirimbas	x	x			
	NE	Aïr et Ténéré	x	x			
	NG	Gadabedji	x	x			
	NG	Odzala		x			
	NG	Omo	x	x			
	RW	Gishwati-Mukura Landscape		x			
	RW	Volcans		x			

Biosphere reserve's roles in climate change projects							
	Country code	Biosphere reserve (n=28)	Leading role	Partner role	Coordi-nating role	Obser-ver role	No role
UNESCO AfriMAB Biosphere Reserves represented	SZ	Lubombo	x	x	x		
	TZ	Jozani - Chwaka Bay			x	x	
	ZA	Cape West Coast					x
	ZA	Cape Winelands		x			
	ZA	Gouritz Cluster	x	x	x	x	
	ZA	Kogelberg			x		
	ZA	Kruger to Canyons	x		x		
	ZA	Magaliesberg	x			x	
	ZA	Marico	x		x		
	ZA	Vhembe	x	x	x	x	
		Total	19	22	14	5	1

Source: Own compilation

Table 26 Funding source origin.

Funding source origins						
	Country code	Biosphere reserves (n=28)	Internatio-nal	Federal/ national	State/ provincial	Local
UNESCO AfriMAB Biosphere Reserves represented	BF	Mare aux hippopotames	x	x	x	
	BJ, BF, NE	WAP	x	x	x	
	BJ, TG	Mono Transboundary	x		x	
	CD	Luki	x			
	CD	Yangambi	x			
	CG	Mount Mulanje	x			x
	CI	Comoé	x	x	x	x
	CI	Taï	x	x	x	x
	CM	Dja	x			x
	GH	Songor				x
	MG	Sahamalaza - Iles Radama	x			
	MG	Tsimanampesotse - Nosy Ve Androka	x	x		
	ML	Boucle du Baoulé	x			x
	MW	Odzala	x	x		x
	MZ	Quirimbas	x	x	x	x
	NE	Aïr et Ténéré	x			
	NG	Gadabedji	x			
	NG	Omo	x	x	x	x
	RW	Gishwati-Mukura Lands-cape	x			x

Funding source origins						
UNESCO AfriMAB Biosphere Reserves represented	Country code	Biosphere reserves (n=28)	Internatio-nal	Federal/ national	State/ provincial	Local
	RW	Volcans	x			x
	SZ	Lubombo	x			x
	TZ	Jozani - Chwaka Bay	x			
	ZA	Cape West Coast			x	x
	ZA	Cape Winelands	x			
	ZA	Gouritz Cluster	x			x
	ZA	Kruger to Canyons	x			x
	ZA	Magaliesberg	x			x
	ZA	Vhembe	x	x	x	x
		Total	**26**	**9**	**9**	**17**

Source: Own compilation

Table 27 Climate change adaptation or mitigation initiatives origin for sustainability.

Climate change adaptation or mitigation initiatives origin for sustainability			
UNESCO MAB level		N	Percent of responses
	Local	25	58%
	National	10	23%
	International	3	7%
	Other	5	12%
	Total (n=43)	43	100%

Source: Own compilation

Table 28 Climate change strategies of AfriMAB.

Country code	Biosphere reserves (n=29)	Creating public awareness/understanding for climate change	Involving new stakeholders including, traditional and indigenous communities from outside the biosphere reserve	Sensitizing decision-makers and donors about climate change	Innovative climate change education for children and students	Long term climate change monitoring	Integrating policies of several government entities/departments/agencies	New interdisciplinary research	Mutual learning among stakeholders about climate change	Climate change planning priorities in the participatory process	Local climate change scenarios/predictions	Generating new knowledge by combining results across projects	Involving political/economic stakeholders in research design	Adapting planning, implementing and evaluation processes of the management plan and its frequency	Total
BF	Mare aux hippopotames	x	x	x	x	x	x	x	x	x	x		x	x	12
BJ, BF, NE	WAP	x	x		x	x	x	x	x	x	x			x	10
BJ, TG	Mono Transboundary	x	x	x		x	x						x		5
CD	Luki	x	x	x	x	x	x	x	x	x	x	x	x	x	13
CD	Yangambi	x	x	x	x	x	x	x	x	x	x	x	x	x	13
CG	Odzala		x							x					2
CI	Comoé	x			x				x						3
CI	Taï	x	x	x	x	x	x	x	x	x	x	x	x	x	13
CM	Dja	x	x	x	x	x	x	x		x	x	x		x	11
GH	Songor	x			x										2

MG	Sahamalaza - Iles Radama	x		x		x									3
MG	Tsimanam-pesotse - Nosy Ve Androka	x	x	x	x										4
ML	Boucle du Baoulé	x	x	x	x	x			x	x				x	8
MZ	Quirimbas	x	x	x	x		x		x	x					7
NE	Aïr et Ténéré	x	x	x	x	x	x	x	x			x	x		10
NG	Gadabedji	x	x	x	x	x	x	x	x			x	x		10
NG	Omo	x	x	x	x		x	x	x						7
RW	Gishwati-Mukura Landscape	x	x	x	x	x	x	x	x		x	x	x		11
RW	Volcans	x	x		x	x	x	x	x		x	x	x		10
SZ	Lubombo	x	x	x	x	x		x	x	x	x	x		x	11
TZ	Jozani - Chwaka Bay	x	x	x	x	x	x	x			x	x	x		10
ZA	Cape West Coast	x		x					x						3
ZA	Cape Winelands	x	x	x		x		x	x						6
ZA	Gouritz Cluster	x	x	x	x		x	x	x	x	x	x	x	x	12
ZA	Kogelberg	x	x	x	x										4
ZA	Kruger to Canyons		x			x	x	x		x	x	x		x	8
ZA	Magaliesberg		x	x											2
ZA	Marico	x			x		x	x							4
ZA	Vhembe	x	x	x	x	x	x			x	x	x			9
	Total	26	24	22	22	18	18	18	17	13	13	13	11	10	

Source: Own compilation

Table 29 Governance system performance measures.

			Relationship classification of UNESCO biosphere reserves and network levels															Resilience of UNESCO biosphere reserves institutional structure					Biosphere reserve ability for long-term implementation of climate change projects.						
			International					National					Regional																
		Biosphere reserves (n=31)	Medium (3)	Medium-low (2)	Medium (3)	Medium-weak (2)	Weak (1)	Strong (5)	Strong-medium (4)	Medium (3)	Medium-weak (2)	Weak (1)	Strong (5)	Strong-medium (4)	Medium (3)	Medium-weak (2)	Weak (1)	High (5)	High-medium (4)	Medium (3)	Medium-low (2)	Low (1)	Strongly agree (5)	Agree (4)	Neutral (3)	Disagree (2)	Strongly disagree (1)	Total	Percentage
UNESCO AfriMAB Biosphere Reserves represented	BF	Mare aux hippopotames					x			x						x				x				x				13	52%
	BJ, BF, NE	WAP			x				x						x					x				x				17	68%
	CI	Comoé		x				x						x					x					x				21	84%
	CI	Taï		x					x					x						x				x				19	76%
	CM	Benoué	x							x				x					x							x		18	72%
	CM	Dja	x						x					x					x						x			20	80%
	CM	Doumba-Rey	x							x				x					x							x		18	72%
	CM	Waza	x							x				x					x							x		18	72%
	ET	Kafa					x					x					x			x					x			9	36%
	ET	Majang Forest					x					x					x			x					x			9	36%
	ET	Sheka					x					x					x			x					x			9	36%
	GW	Boloma-Bijagós			x			x						x				x							x			20	80%
	ML	Boucle du Baoulé	x					x						x						x			x					22	88%
	MW	Mount Mulanje			x				x					x					x					x				19	76%

UNESCO AfriMAB Biosphere Reserves represented	NE	Aïr et Ténéré		x						x				x						x				x				18	72%
	NE	Gadabedji		x						x				x						x				x				18	72%
	NG	Omo	x					x					x					x					x					25	100%
	RW	Gishwati-Mu-kura Lands-cape					x				x						x		x						x			11	44%
	RW	Volcans			x					x						x				x					x			14	56%
	SZ	Lubombo	x					x					x								x				x			20	80%
	TZ	East Usambara						x												x				x				12	48%
	TZ	Jozani - Chwaka Bay						x												x				x				12	48%
	TZ	Lake Manyara						x												x				x				12	48%
	TZ	Serengeti-Ngorongoro						x												x				x				12	48%
	ZA	Cape West Coast		x							x			x							x					x		14	56%
	ZA	Gouritz Cluster	x							x			x							x			x					21	84%
	ZA	Kogelberg		x								x			x							x		x				13	52%
	ZA	Kruger to Canyons				x					x					x			x					x				14	56%
	ZA	Magaliesberg		x					x					x						x					x			18	72%
	ZA	Vhembe	x					x					x								x			x				21	84%
	ZA	Waterberg		x					x				x							x				x				20	80%
		Total	9	8	4	1	5	10	6	8	3	4	5	13	2	3	4	2	8	17	3	1	3	15	9	4	0		

Source: Own compilation

Table 30 MAB stakeholders responsible for climate change mitigation and adaptation projects.

UNESCO MAB stakeholders responsible for climate change mitigation and adaptation projects														
UNESCO AfriMAB Biosphere Reserves represented	Country code	Biosphere reserve (n=33)	International Coordinating Council	Thematic/regional network administration	Thematic/regional network membership	National committee administration	National committee membership	Biosphere Reserve coordinator/chairperson	Formal conservation area management	Government organisation	Biosphere reserve board member/director	Closely affiliated company	Research	Consultation/advisory	Total
	BF	Mare aux hippopotames				x					x				2
	BJ, TG	Mono Transboundary				x					x				2
	BJ, BF, NE	WAP	x	x		x	x	x	x	x	x				8
	CD	Yangambi	x			x	x	x		x	x	x	x	x	9
	CI	Comoé				x				x	x		x		4
	CI	Taï				x	x	x		x	x		x		6
	CM	Benoué				x		x	x						3
	CM	Dja				x	x	x	x			x	x		6
	CM	Doumba-Rey				x	x	x	x						4
	CM	Waza				x	x	x							3
	ET	Kafa					x	x		x	x	x			5
	ET	Majang Forest					x	x		x	x	x			5
	ET	Sheka					x	x		x	x	x			5

UNESCO AfriMAB Biosphere Reserves represented	GW	Boloma-Bijagós			x										1
	ML	Boucle du Baoulé		x		x	x	x	x	x	x	x	x	x	10
	MW	Mount Mulanje					x	x		x		x	x		5
	NE	Aïr et Ténéré		x		x		x		x					4
	NE	Gadabedji		x		x	x	x		x					5
	NG	Omo								x			x		2
	RW	Gishwati-Mukura Landscape					x	x	x	x			x		5
	RW	Volcans				x	x	x	x	x			x	x	7
	SZ	Lubombo	x	x	x	x	x	x	x	x	x	x	x	x	12
	TZ	East Usambara	x			x				x	x		x		5
	TZ	Jozani - Chwaka Bay	x			x				x	x		x		5
	TZ	Lake Manyara	x			x				x	x		x		5
	TZ	Serengeti-Ngorongoro	x			x				x	x		x		5
	ZA	Cape West Coast		x	x		x	x	x	x	x	x			8
	ZA	Gouritz Cluster									x				1
	ZA	Kogelberg					x	x							2
	ZA	Kruger to Canyons		x			x	x		x					4
	ZA	Magalies-berg					x	x	x	x	x	x	x		7
	ZA	Vhembe					x	x	x	x	x	x	x		7
	ZA	Waterberg	x	x	x	x	x	x	x	x	x	x	x	x	12
		Total	8	8	4	20	21	23	12	24	20	12	17	5	

Source: Own compilation

Table 31 AfriMAB stakeholders who should be responsible for research.

	UNESCO MAB stakeholders that should be responsible for research														
UNESCO AfriMAB Biosphere Reserves represented	Country Code	Biosphere reserves (n=33)	International Coordinating Council	Thematic/regional network administration	Thematic/regional network membership	National committee administration	National committee membership	Biosphere Reserve coordinator/chairperson	Formal conservation area management	Government organisation	Biosphere reserve board member/director	Closely affiliated company to the biosphere reserve	Research	Consultation/advisory	Total
	BF	Mare aux hippopotames					x	x			x		x	x	5
	BJ, TG	Mono Transboundary			x								x	x	3
	BJ, BF, NE	WAP					x	x	x	x	x		x	x	7
	CD	Yangambi	x	x		x	x	x		x	x	x	x	x	10
	CI	Comoé					x			x	x		x	x	5
	CI	Taï		x		x	x	x		x	x		x	x	8
	CM	Benoué				x		x	x	x					4
	CM	Dja				x		x	x	x			x	x	6
	CM	Doumba-Rey				x		x	x	x					4
	CM	Waza				x		x	x	x					4
	ET	Kafa	x					x	x		x	x	x	x	7
	ET	Majang Forest	x					x	x		x	x	x	x	7

UNESCO AfriMAB Biosphere Reserves represented	ET	Sheka	x					x	x		x	x	x	x	7
	GW	Boloma-Bijagós	x	x			x				x				4
	ML	Boucle du Baoulé		x	x	x	x	x	x	x	x	x	x	x	11
	MW	Mount Mulanje		x			x	x		x			x	x	6
	NE	Aïr et Ténéré								x			x	x	3
	NE	Gadabedji								x			x	x	3
	NG	Omo					x			x			x	x	4
	RW	Gishwati-Mukura Landscape					x	x	x	x			x	x	6
	RW	Volcans					x	x	x	x		x	x	x	7
	SZ	Lubombo					x						x	x	3
	TZ	East Usambara	x							x			x	x	4
	TZ	Jozani - Chwaka Bay	x							x			x	x	4
	TZ	Lake Manyara	x							x			x	x	4
	TZ	Serengeti-Ngorongoro	x							x			x	x	4
	ZA	Cape West Coast	x	x	x			x	x	x	x		x	x	9
	ZA	Gouritz Cluster									x				1
	ZA	Kogelberg							x						1
	ZA	Kruger to Canyons											x	x	2
	ZA	Magaliesberg											x		1
	ZA	Vhembe	x			x		x	x	x	x	x	x	x	9
	ZA	Waterberg								x	x	x	x	x	5
		Total	11	6	3	8	12	17	14	22	14	8	26	26	

Source: Own compilation

Table 32 Organisations working within the AfriMAB Biosphere Reserves.

Country (n=12)	Organisations working within Biosphere reserves
Burkina Faso	1) RBT-WAP/GIC-WAP implemented by GIZ, 2) IUCN/PAPBIO, 3) Ministry of the Environment (OUEDRAOGO amadé, Yaméogo Souleymane, Compaoré Prospert DOAMBA Bénoît).
Cameroon	1) Congo Bassin Institut (CBI), 2) SudCam Hévéa SA, 3) PALLISCO – LAGOUTE, 4) CAMINEX - LEKINA NGA, 5) HydroMeking, 6) PA-DIDJA/MINEPAT - TALA TALA, 7) FANKEM, 8) Zoological Society of London (ZSL), 9) African Wildlife Foundation (AWF).
Democratic Republic of Congo	1) School of Integrated Management of Tropical Forests and Lands (ERAIFT), 2) European Union, 3) Belgium, 4) Africa Museum, 5) the World Bank, 6) the Universities of Gent and ULB, 7) CIFOR, 8) RMCA, 9) WWF
Côte d'Ivoire	1) GIZ, 2) the Ivorian Office of Parks and Reserves (OIPR) 3) Swiss Centre for Scientific Research (CSRS).
Eswatini	1) COSPE (Italian cooperation agency), 2) World Vision, 3) the University of Swaziland, 4) Lubombo Conservancy, 5) Trans-Frontier Conservation Area, 6) the Eswatini National Trust Commission, 7) Charuma Conservation Initiative, 8) Mbuluzi Game Reserve, 9) Big Game Parks.
Malawi	1) NGO's
Mozambique	1) WWF, 2) Unilurio University of Mozambique
Nigeria	1) Nigeria Conservation Foundation
Rwanda	1) NGO's
South Africa	1) WATSAN Africa, 2) Cape Nature, 3) University of Stellenbosch, 4) WWF, 5) Greater Kruger Strategic Development Programme.
Tanzania	1) Terra globe, 2) IFAD.
Togo	1) REDISSE-Togo, 2) Ministry of Health, 3) Ministry of Livestock, 4) Ministry of the Environment.

Source: Own compilation

Table 33 Climate change project activity of organisations independent from the biosphere reserve in different zones.

Climate change project activity of organisations independent from the biosphere reserve in different zones														
			Core				Buffer				Transition			
	Country code	Biosphere reserves (n=31)	Very active (3)	Moderately active (2)	Not very active (1)	Inactive (0)	Very active (3)	Moderately active (2)	Not very active (1)	Inactive (0)	Very active (3)	Moderately active (2)	Not very active (1)	Inactive (0)
UNESCO AfriMAB Biosphere Reserves represented	BF	Mare aux hippopotames		x				x				x		
	BJ, BF, NE	WAP		x				x				x		
	BJ, TG	Mono Transboundary			x				x			x		
	CD	Luki	x					x				x		
	CD	Yangambi	x					x				x		
	CG	Odzala	x					x						
	CI	Comoé	x					x				x		
	CI	Taï	x					x				x		
	CM	Dja	x					x				x		
	GH	Songor		x				x					x	
	MG	Sahamalaza - Iles Radama	x					x					x	
	MG	Tsimanampesotse - Nosy Ve Androka	x					x			x			
	ML	Boucle du Baoulé		x				x			x			
	MW	Mount Mulanje	x				x				x			
	MZ	Quirimbas		x				x				x		
	NE	Aïr et Ténéré		x				x					x	
	NE	Gadabedji		x				x					x	

UNESCO AfriMAB Biosphere Reserves represented	NG	Omo		x				x					x	
	NG	Odzala	x					x						
	RW	Gishwati-Mukura Landscape		x				x				x		
	RW	Volcans		x				x				x		
	SZ	Lubombo		x			x					x		
	TZ	Jozani - Chwaka Bay		x				x				x		
	ZA	Cape West Coast			x				x				x	
	ZA	Cape Winelands			x				x				x	
	ZA	Gouritz Cluster		x				x			x			
	ZA	Kogelberg			x			x					x	
	ZA	Kruger to Canyons		x				x				x		
	ZA	Magaliesberg		x				x					x	
	ZA	Marico			x				x				x	
	ZA	Vhembe		x				x					x	
		Total	10	16	5	0	2	25	4	0	4	14	11	0

Source: Own compilation

Table 34 Respondents' awareness of the impact of climate change.

Responses	N	Percentage
Yes	17	32.1%
Partly	27	50.9%
No	9	17.0%
Total (n=53)	53	100.0%

Source: Own compilation

Table 35 Actor performance measures.

	Country code	Biosphere reserve (n=32)	Livelihood resilience to climate change					Relationship strength between biosphere reserve, local populations, businesses, and institutions					Local populations, businesses and institutions are aware of the impact of climate change					Companies are aware of the impact of temperature increase					Total	Percentage
			High (5)	High-medium (4)	Medium (3)	Medium-low (2)	Low (1)	Strong (5)	Strong-medium (4)	Medium (3)	Medium-weak (2)	Weak (1)	Strongly agree (5)	Agree (4)	Neutral (3)	Disagree (2)	Strongly disagree (1)	Strongly agree (5)	Agree (4)	Neutral (3)	Disagree (2)	Strongly disagree (1)		
UNESCO AfriMAB Biosphere Reserves represented	BF	Mare aux hippopotames			x			x					x					x					18	90%
	BJ, BF, NE	WAP			x				x					x						x			14	70%
	CD	Yangambi		x				x					x					x					19	95%
	CI	Comoé			x				x						x				x				14	70%
	CI	Taï			x				x						x					x			13	65%
	CM	Benoué			x							x					x			x			8	40%
	CM	Dja			x						x				x					x			11	55%
	CM	Doumba-Rey			x							x					x			x			8	40%

UNESCO AfriMAB Biosphere Reserves represented	CM	Waza			x							x					x			x			8	40%
	ET	Kafa			x						x				x					x			11	55%
	BF	Majang Forest		x							x				x					x			12	60%
	BJ, BF, NE	Sheka		x							x				x					x			12	60%
	GW	Boloma-Bijagós		x					x					x					x				16	80%
	ML	Boucle du Baoulé		x					x				x						x				17	85%
	MW	Mount Mulanje	x							x				x					x				16	80%
	NE	Aïr et Ténéré			x				x							x			x				13	65%
	NE	Gadabedji			x				x							x			x				13	65%
	NG	Omo	x					x					x					x					20	100%
	RW	Gishwati-Mukura Land-scape				x				x						x					x		9	45%
	RW	Volcans			x					x				x							x		12	60%
	SZ	Lubombo				x				x					x					x			11	55%

UNESCO AfriMAB Biosphere Reserves represented	TZ	East Usambara		x					x					x						x			15	75%
	TZ	Jozani - Chwaka Bay		x					x					x						x			15	75%
	TZ	Lake Manyara		x					x					x						x			15	75%
	TZ	Serengeti-Ngorongoro		x					x					x						x			15	75%
	ZA	Cape West Coast				x				x				x					x				13	65%
	ZA	Gouritz Cluster	x					x						x					x				18	90%
	ZA	Kogelberg			x					x					x						x		11	55%
	ZA	Kruger to Canyons				x				x				x					x				13	65%
	ZA	Magaliesberg				x				x						x				x			10	50%
	ZA	Vhembe				x			x					x						x			13	65%
	ZA	Waterberg	x								x			x					x				15	75%
		Total	4	9	13	6	0	4	12	8	5	3	4	13	8	4	3	3	10	16	3	0		

Source: Own compilation

Table 36 Cumulative performance score of the resource system, resource units, governance systems and actors.

Performance measures cumulative score								
	Country code	Biosphere reserves (n=32)	Resource system (15)	Resource units (7)	Governance system (25)	Actors (20)	Total: (_/67)	Percentage
UNESCO AfriMAB Biosphere Reserves represented	BF	Mare aux hippopotames	11	6	13	18	48	72%
	BJ, BF, NE	WAP	10	4	17	14	45	67%
	CD	Yangambi	12	1		19	32	48%
	CI	Comoé	11	3	21	14	49	73%
	CI	Taï	10	4	19	13	46	69%
	CM	Benoué	8	5	18	8	39	58%
	CM	Dja	9	5	20	11	45	67%
	CM	Doumba-Rey	8	5	18	8	39	58%
	CM	Waza	8	5	18	8	39	58%
	ET	Kafa	11	4	9	11	35	52%
	ET	Majang Forest	11	5	9	12	37	55%
	ET	Sheka	11	4	9	12	36	54%
	GW	Boloma-Bijagós	9		20	16	45	67%
	ML	Boucle du Baoulé	10	7	22	17	56	84%
	MW	Mount Mulanje	8	4	19	16	47	70%
	NE	Aïr et Ténéré	5	1	18	13	37	55%
	NE	Gadabedji	5	1	18	13	37	55%
	NG	Omo	12	7	25	20	64	96%
	RW	Gishwati-Mukura Landscape	6	3	11	9	29	43%

UNESCO AfriMAB Biosphere Reserves represented	RW	Volcans	8	5	14	12	39	58%
	SZ	Lubombo	6	3	20	11	40	60%
	TZ	East Usam-bara	10	5	12	15	42	63%
	TZ	Jozani - Chwaka Bay	10	5	12	15	42	63%
	TZ	Lake Manyara	10	5	12	15	42	63%
	TZ	Serengeti-Ngorongoro	10	5	12	15	42	63%
	ZA	Cape West Coast	7	3	14	13	37	55%
	ZA	Gouritz Cluster	9	3	21	18	51	76%
	ZA	Kogelberg	9	2	13	11	35	52%
	ZA	Kruger to Canyons	9	5	14	13	41	61%
	ZA	Magaliesberg	8	4	18	10	40	60%
	ZA	Vhembe	8	2	21	13	44	66%
	ZA	Waterberg	11	4	20	15	50	75%
		Total average	9	4	16	13	40	60%

Source: Own compilation

Table 37 Comparison of top three best and worst performing AfriMAB Biosphere Reserves.

Indicator elements	Best performing biosphere reserves			Worst performing biosphere reserves		
Biosphere reserve	Omo	Boucle du Baoulé	Gouritz Cluster	Gishwati-Mukura Landscape	Kafa	Kogelberg
Country	Nigeria	Mali	South Africa	Rwanda	Ethiopia	South Africa
Region	West Africa	West Africa	Southern Africa	East Africa	East Africa	Southern Africa
Country income classification	Lower middle	Low	Lower middle	Low	Low	Lower middle
Age	1977	1982	2015	2020	2010	1998
Major biome	Tropical forest	Savannah	Mountain	Mountain, tropical forest	Mountain, tropical forest	Mountain, marine coastal areas
Number of projects	NA	2	4	2	NA	0
First climate change project	NA	2017	2012	2020	2017	NA
Zone focus	All	Transition	All	All	NA	Transition
Climate change impact study	Yes	Yes	No	NA	NA	NA
CO_2 assessment	Yes	No	Yes	No	NA	No
Project categories	Agriculture & livelihoods, forest conservation, natural resource protection	Education	Agriculture & livelihoods, community, natural resource protection	Education, community, natural resource protection	NA	NA

Indicator elements	Best performing biosphere reserves			Worst performing biosphere reserves		
Estimated cost per km^2 project	NA	USD 18.73 km^2	USD 10.72 km^2	NA	NA	NA
Estimated duration project	NA	2	4	1	NA	1
Institutionalisation	58%	50%	92%	83%	NA	25%
Role	Leading & coordinating	Partner & coordinating	All roles	Partner	NA	Coordinating
Funding sources	All	International & local	International & local	International & local	NA	NA
Responsible stakeholder for projects	Government organisations & research	All except ICC and thematic network members	Biosphere reserve board member	National committee, CEO, formal conservation, government organisation & research	National committee, CEO, government organisation, biosphere reserve board member, closely affiliated company, research & advisors	National committee & CEO
Research stakeholders	National committee, government organisation, research & advisors	All except ICC	Biosphere reserve board member	National committee and members, formal conservation, government organisation, research & advisory	ICC, CEO, formal conservation, biosphere reserve board member, closely affiliated company, research & advisors	Formal conservation management
Other organisation activity	Moderately active core and buffer, not very active in transition	Moderately active core and buffer, very active in transition	Moderately active core and buffer, very active in transition	Moderately active all zones	NA	Not very active in the core and transition, but very active in buffer

Source: Own compilation

Table 38 Efforts to limiting global temperature to below 2°C from least to most effort.

	Country code	Biosphere reserves (n=30)	Ranking of efforts to limiting global temperature to below 2°C from least to most effort									
			1	2	3	4	5	6	7	8	9	10
UNESCO AfriMAB Biosphere Reserves represented	BF	Mare aux hippopotames					x					
	BJ, TG	Mono Transboundary			x							
	BJ, BF, NE	WAP				x						
	CD	Yangambi						x				
	NG	Omo							x			
	NG	Odzala							x			
	CG	Luki						x				
	CI	Comoé							x			
	CI	Taï							x			
	CM	Dja							x			
	MG	Sahamalaza - Iles Radama										x
	MG	Tsimanampesotse - Nosy Ve Androka										x
	ML	Boucle du Baoulé				x						
	MW	Mount Mulanje			x							
	MZ	Quirimbas							x			
	NE	Aïr et Ténéré								x		
	GH	Songor						x				
	NE	Gadabedji								x		
	RW	Gishwati-Mukura Landscape							x			

UNESCO AfriMAB Biosphere Reserves represented	RW	Volcans							x			
	TZ	Jozani - Chwaka Bay	x									
	SZ	Lubombo					x					
	ZA	Cape Wine-lands	x									
	ZA	Cape West Coast			x							
	ZA	Marico				x						
	ZA	Gouritz Clus-ter					x					
	ZA	Kogelberg				x						
	ZA	Kruger to Canyons			x							
	ZA	Magaliesberg						x				
	ZA	Vhembe						x				
		Total	2	0	4	4	3	5	8	2	0	2

Source: Own compilation

Table 39 Scenarios into which organisation's efforts fit in the first questionnaire.

Scenarios in which the organisation's efforts fit					
	Country code	Biosphere reserve (n=56)	Scenario 1: Business as usual	Scenario 2: Midd-ling target	Scenario 3: Ideal target
UNESCO AfriMAB Biosphere Reserves represented	BF	Mare aux hippopotames		2	1
	BJ, TG	Mono Transboundary		1	
	BJ, BF, NE	WAP		4	1
	CD	Yangambi		2	
	CG	Luki		2	
	CG	Odzala		2	
	CI	Taï		2	2
	CI	Comoé	2	2	
	CM	Dja	1	2	

UNESCO AfriMAB Biosphere Reserves represented	GH	Songor		1	
	MG	Tsimanampesotse - Nosy Ve Androka			
	ML	Boucle du Baoulé		2	
	MW	Mount Mulanje		1	
	MZ	Quirimbas	1	1	1
	NE	Aïr et Ténéré		1	
	NE	Gadabedji		1	
	NG	Omo	1		1
	RW	Gishwati-Mukura Lands-cape		1	
	RW	Volcans		1	
	SZ	Lubombo		1	1
	TZ	Jozani - Chwaka Bay	1		
	ZA	Cape West Coast		1	
	ZA	Cape Winelands		1	
	ZA	Gouritz Cluster		1	1
	ZA	Kogelberg		2	
	ZA	Kruger to Canyons		1	
	ZA	Magaliesberg		1	
	ZA	Marico		1	
	ZA	Vhembe		1	
		Total respondents	6	38	8

Source: Own compilation

Table 40 Scenarios into which organisation's efforts fit in the second questionnaire.

Scenarios in which the organisation's efforts fit					
UNESCO AfriMAB Biosphere Reserves numbers of representatives	Country code	Biosphere reserves (n=25)	Scenario 1: Business as usual where the average global temperature could increase by at least 4°C	Scenario 2: Middling where the average global temperature could increase by 2°C	Scenario 3: Ideal target where the average global temperature could increase by 1.5°C
	BF	Mare aux hippopotames			1
	BJ, BF, NE	WAP	1	1	2
	CD	Yangambi			1
	CI	Comoé			2
	CI	Taï		1	3
	CM	Benoué			1
	CM	Dja			2
	CM	Doumba-Rey			1
	CM	Waza			1
	ML	Boucle du Baoulé			1
	MW	Mount Mulanje		1	
	NE	Aïr et Ténéré			1
	NE	Gadabedji			1
	NG	Omo	1		
	RW	Volcans			1
	SZ	Lubombo			1
	TZ	East Usambara			1
	TZ	Jozani - Chwaka Bay			1
	TZ	Lake Manyara			1
	TZ	Serengeti-Ngorongoro			1
	ZA	Cape West Coast			1

	ZA	Gouritz Cluster		1	
	ZA	Kogelberg		1	
	ZA	Kruger to Canyons		1	
	ZA	Magaliesberg		1	
	ZA	Vhembe			1
	ZA	Waterberg		1	
		Total respondents	2	8	15

Source: Own compilation

Table 41 Factors determining the pursuit of climate change projects.

Country code	Biosphere reserves (n=29)	Factors determining the pursuit of climate change projects				
BF	Mare aux hippopotames	Funding	Publish successes and failures			
BJ, TG	Mono Transboundary	Funding	Priorities and interest of technical and financial partners	Increasing effects of climate change requiring urgent action	Low resilience of local communities	
BJ, BF, NE	WAP	Funding	Project design	Publish successes and failures	Site safety	Research monitoring and evaluation.
CD	Yangambi	Organisation mandate	Capacity building of biosphere reserve management structure	Stakeholder involvement	Opportunities for mobilising additional resources	

CG	Odzala	Research, monitoring and evaluation	Valorisation of carbon credits			
CG	Luki	Organisation mandate	Capacity building of biosphere reserve management structure	Stakeholder involvement	Opportunities for mobilising additional resources	
CI	Taï	Funding	Climate change impact	Stakeholder involvement	Technical support	Capacity of management structure
CI	Comoé	Funding	Project impact	Stakeholder involvement	Technical support	Capacity building of the biosphere reserve management structure
CM	Dja	Commitment and determination of stakeholders	Sustainable funding	Technical support	Project benefits to stakeholders	
GH	Songor	Site-specific projects	Project availability	Funding		
MG	Tsimanampesotse - Nosy Ve Androka	Funding	Changing attitudes of local people			
ML	Boucle du Baoulé	Stakeholder involvement	Projects developed in the villages bordering the Reserve	Climate Change Education in Schools	Temperature variability	Projects addressing climate change related challenges
MW	Mount Mulanje	Funding	Technical support	Climate information		
MZ	Quirimbas	Changing attitudes of local people	Government involvement	Stakeholder involvement		

NE	Aïr et Ténéré	Organisation mandate	Recovery of degraded land	Management and Development of Protected Areas	Low resilience of local communities	
NE	Gadabedji	Organisation mandate	Recovery of degraded land	Management and Development of Protected Areas	Low resilience of local communities	
NG	Omo	Organisation mandate	Funding			
RW	Gishwati-Mukura Landscape	Expertise	Stakeholder involvement	Local knowledge of climate change		
RW	Volcans	Expertise	Stakeholder involvement	Local knowledge of climate change		
SZ	Lubombo	Funding	Expertise			
TZ	Jozani - Chwaka Bay	Changing attitudes of local people				
ZA	Cape Winelands	Funding	Government involvement	University involvement		
ZA	Cape West Coast	Funding	Water scarcity			
ZA	Marico	Funding	Project Relevance and feasibility			
ZA	Gouritz Cluster	Projects addressing climate change related challenges	Presence of local champion	Project benefits		

ZA	Kogelberg	Capacity building of the biosphere reserve management	Stakeholder involvement			
ZA	Kruger to Canyons	Funding				
ZA	Magaliesberg	Funding	Stakeholder involvement			
ZA	Vhembe	Funding	Stakeholder involvement			

Source: Own compilation

Table 42 Suggestions to limit the global temperature increase.

Country code	BF	BJ, BF, NE	CI	CM	CM	CM	ET	ET	ML	MW	NE	NG	RW	SZ	TZ	TZ	TZ	TZ	ZA	ZA	ZA	ZA	ZA	
Biosphere reserves (n=41)	Mare aux hippopotames	WAP	Comoé	Benoué	Dja	Doumba-Rey	Kafa	Majang Forest	Boucle du Baoulé	Mount Mulanje	Aïr et Ténéré	Omo	Gishwati-Mukura Landscape	Lubombo	East Usambara	Jozani - Chwaka Bay	Lake Manyara	Serengeti-Ngorongoro	Cape West Coast	Gouritz Cluster	Kogelberg	Kruger to Canyons	Magaliesberg	Total elements
Social, economic, and political settings																								
More direct involvement and intervention from UNESCO into the 25-50 hotspot biosphere reserves and showcase this in a worldwide awareness project. UNESCO does not do enough.	x	x	x	x	x	x			x		x		x		x	x	x	x						13
Orientation of national decision-makers on the challenges of global warming.	x	x		x	x	x			x		x		x	x	x	x	x	x						13
The network could be more active in each country to promote meetings, workshops, and actions to elevate this issue.		x		x	x	x			x	x			x	x	x	x	x	x						14

Governance systems																							
Building capacity of institutions in climate policy review and analysis.	x	x	x	x	x	x	x	x	x		x	x	x	x								x		15
By promoting the exchange of experience between biosphere reserves to share success stories on the issue of climate change on a local, regional, and international level.	x	x	x	x	x	x	x	x	x					x					x				x	12
Intensify and support experience-sharing meetings at local, regional, and international levels.	x	x	x		x				x		x		x	x										8
The government should also provide resources to support the initiatives of Indigenous people.	x	x					x	x	x		x	x	x	x										9
Capacity building of biosphere reserves to maintain their role as learning centres for sustainable development by perfectly reconciling protection and development.	x	x	x	x	x	x	x	x	x	x	x	x	x	x					x		x		x	17
Capacity building for biosphere reserve managers.	x	x	x		x				x	x	x	x	x	x						x				12

Closer collaboration at all levels.	x	x		x	x	x	x	x	x			x	x	x					x		x			13
Strengthening the technical capacities of managers.	x	x			x		x	x	x	x	x	x	x	x							x		x	13
Orientation of national decision-makers on the challenges of global warming.	x	x		x	x	x			x		x		x	x	x	x	x	x						13
The network could be more active in each country to promote meetings, workshops, and actions to elevate this issue.		x		x	x	x			x	x			x	x	x	x	x	x						14
Resource system																								
Strengthening the monitoring and conservation of core areas of biosphere reserves.	x	x	x	x	x	x	x	x	x		x		x	x	x	x	x	x	x				x	18
Improving conservation of the buffer zone and transition of the reserve.	x	x	x	x	x	x	x	x	x		x		x	x	x	x	x	x	x		x		x	19
Enhancing the conservation of biodiversity including forest resources.	x	x					x	x	x		x			x	x	x	x	x					x	12

Strengthening the supervision of actions carried out in the buffer zones.	x	x					x	x	x		X			x					x		x		x	10
Developing action in favour of the populations in zones bordering the protected areas.	x	x	x	x	x	x			x	x	x	x	x	x	x	x	x	x	x				x	18
Integration of the climate change issue into feasible development projects in the transition zone.	x	x		x	x	x	x	x	x	x	x		x	x					x				x	14
Designating more biosphere reserves and engaging in forest restoration of degraded lands, especially in the transition zones of the biosphere reserves.	x	x	x		x					x	x			x										7
Contributing to the reforestation and enrichment of agricultural areas with forest species of interest (through agroforestry methods) in the transition zone.	x	x	x		x		x	x	x	x		x	x	x						x		x		13
There should be a central information portal on climate change so all the efforts by individual biosphere reserves can be coordinated and shared.	x	x			x				x					x	x	x	x	x	x				x	11

Evaluation of the amount of CO_2 sequestered by tropical forest biosphere reserves to proceed with financial compensation. This financial windfall could be used for activities compatible with the conservation of natural resources for the benefit of locals.	x	x	x	x	x	x	x	x	x	x	x			x	x	x	x	x						16
Resource units																								
Financially support the biosphere reserve to implement projects aimed at mitigating climate change.	x	x	x	x	x	x	x	x	x		x	x		x	x	x	x	x	x		x		x	19
Become more involved with the acquisition of funding for biosphere reserves that are reaching their goals.	x	x	x				x	x	x	x	x	x	x	x	x	x	x	x	x	x	x	x	x	20
Strengthening the supervision of actions carried out in the buffer zones.	x	x					x	x	x		x			x					x		x		x	10
Developing action in favour of the populations in zones bordering the protected areas.	x	x	x	x	x	x			x	x	x	x	x	x	x	x	x	x	x				x	18

Enabling the conceptualization of bankable projects.	x	x			x				x	x			x	x									x	8
Lobbying for the effective payment of carbon credits.	x	x	x	x	x	x			x					x							x			9
There should be projects that focus on quantifying the carbon footprint of the biosphere reserves, followed by projects that aim to reduce the carbon footprint, or promote activities that lead to a lower carbon footprint.	x	x	x						x	x	x	x	x	x						x		x		11
When the MAB Secretariat or other donors make funds available to MAB National Committees for restoration, monitoring and training activities, people will be more motivated to preserve the resources.	x	x		x	x	x	x	x	x			x	x	x	x	x	x	x						15
Actors																								
Use of a network of experts in organizing country-level training/seminars around climate change topics: global temperature increases to below 1.5^{0} C by 2050.		x	x	x	x	x	x	x	x		x		x	x	x	x	x	x						15

Education should be enhanced particularly for industries emitting large quantities of CO_2 (to pursue mitigating factors and an increase in the climate change adaptation projects) and at a local level to assist communities to cope with the effects of climate.	x	x	x		x				x		X		x	x	x	x	x	x	x		x			14
Raise awareness of biosphere reserves among the public and decision-makers.	x	x	x	x	x	x	x	x	x		x	x	x	x	x	x	x	x	x		x			19
Raise people's awareness of the challenges and benefits of working on resilience to climate change.	x	x			x		x	x	x	x	x	x	x	x					x		x		x	14
Involve local partners in the restoration of degraded areas.	x	x	x	x	x	x			x	x	x		x	x					x		x		x	14
Increase the participation of local communities living near biosphere reserves in the management and conservation of natural resources and encourage them to reduce their CO_2 emissions and promote sustainable agriculture.	x	x	x	x	x	x	x	x	x	x	x		x	x	x	x	x	x	x				x	19

Promoting the principles of prevention, precaution, and adaptation.	x	x			x				x					x	x	x	x	x					x	10
Support the development of project finances and the search for partners.	x	x	x	x	x	x	x	x	x	x		x	x	x						x		x	x	16
Outcomes																								
Reducing population growth should be a priority while improving livelihoods by increasing the quality of education and raising awareness of climate change issues.		x					x	x		x	x		x	x	x	x	x	x	x		x			13
The emission of carbon dioxide into the environment must be reduced and avoided.	x	x			x		x	x	x		x		x	x					x					10
Biosphere reserves should serve as field laboratories for testing practical solutions to forest ecosystem management problems.	x	x	x	x	x	x			x		x	x	x	x									x	12
UNESCO has major assets that could be released to help plan activities to reconcile conservation and sustainable development.	x	x		x	x	x			x		x	x	x	x	x	x	x	x						14

Create opportunities to promote research on climate change in the biosphere reserves.	x	x	x		x		x	x	X	x	x		x	x									x	12
Develop a green economy in biosphere reserves by creating alternative livelihood projects.	x	x	x	x	x	x			x	x	x		x	x						x	x	x	x	15
Create environmental education clubs and introduce environmental education in primary schools.	x	x			x				x	x	x		x	x	x	x	x	x			x		x	14
Implement biodiversity-related programmes and climate change mitigation projects.	x	x	x						x	x	x	x	x	x	x	x	x	x						13
Enforce fines on polluters.	x	x			x				x		x			x										6
Empower youth organizations.	x	x	x								x		x	x	x	x	x	x						10
Support renewable energy and reduce coal energy.	x	x	x		x				x	x	x			x										8

More intensive food forest approach to smallholder farming.	x	x					x	x	x	x			x	x										8
Prioritise secure large-scale tree planting.	x	x		x	x	x			x	x		x	x	x										10
Total elements per bio-sphere reserve (out of 54)	49	52	29	27	42	27	26	26	49	26	39	20	40	51	25	25	25	25	20	6	16	6	24	

Source: Own compilation